THE NEW STANDARDIZATION

Keystone of Continuous Improvement in Manufacturing

THE NEW STANDARDIZATION

Keystone of Continuous Improvement in Manufacturing

Shigehiro Nakamura

Foreword by
Mark O'Brien

Publisher's Message by
Norman Bodek

Productivity Press
Portland, Oregon Norwalk, Connecticut

Productivity Press
P.O. Box 13390
Portland, OR 97213
United States of America
Telephone: (503) 235-0600
Telefax: (503) 235-0909

Cover design by Gary Ragaglia
Printed and bound by Edwards Brothers in the United States of America
Printed on acid-free paper

Nakamura, Shigehiro.
[QCD kakushin no tame no genba no hyōjunka gijutsu. English]
The new standardization, keystone of continuous improvement in manufacturing/ Shigehiro Nakamura; foreword by Mark O'Brien; [translated by Bruce Talbot].
p. cm.
Includes index.
ISBN 1-56327-039-0
1. Standardization. 2. Production engineering. I. Title.
T59.N35 1993
670.42'0218—dc20 93-14723
CIP

93 94 95 10 9 8 7 6 5 4 3 2

Contents

Publisher's Message ix
Foreword xiii
Preface xv

1 Standardization for a New Era **1**
Errors in the Conventional Approach to Standardization 1
Standards Indicate the Current Level of Production Technology 5
Standardization for the FA/CIM Era 9
Standards Are Made to Be Improved 13

2 Making Quality Products Easily, Inexpensively, and Quickly **19**
Standards from a Manufacturing Perspective 19
Standards Help Solve Work-Site Problems 23
Good Standards Manuals Have Clear Control Points 29
Standardization from the User's Perspective 32

3 Changing Your Approach to Standardization **37**
Good Methods for Making and Checking Standards 37
Skillful Rule-Making for Responses to Abnormalities 41
Charting a Straight Course Toward Higher Standards 45

4 Types of Standards and How They Are Made **53**
Written Standards 53
Standardized Operation Instructions 57
Methods for Extracting Only the Data Needed, Only When Needed 63

Essential Standardization for New-Employee Training 66
Standards Lighten the Manager's Burden 72

5 Standardization for Quality Control 77
What Is Quality? 77
Does It Take Experience to Produce Good Quality? 78
Determining Customer Needs to Achieve Zero Complaints 83
Creating Distinctive Quality 86
Standardization of Problem Analysis 89
Case Studies: How to Analyze Problems 91
Standardization of Know-How 96
Approach to Zero Defects 108

6 Standardization for Equipment Management 115
We Have an Inspection Checklist — Why Don't Breakdowns Decrease? 115
Standardization for Breakdown Analysis 122
Establishing a System and Procedure for Reducing Breakdowns 132
Using Maintenance Check Sheets for Planning New Equipment Installation and Shutdown Maintenance 138
Standardization Is Indispensable for Unassisted Operation 144
Moving Toward Factory Automation and Flexible Manufacturing Systems 151

7 Standardization for Operation Management 163
Problem-Consciousness and the Pioneers of Operation Analysis 163
Practical Techniques for Developing and Displaying Operation Standards 171
Problem-Consciousness Comes Before Methods 183
Discovering the Advantages of a Standard Time System 187
Automating and Upgrading Standardization 197

8 Standardization for Production Management 207
A Change in Production Management Objectives 207
A Full-Fledged Campaign to Shorten Production Lead Time 215
Caution Points Regarding Computerization 218

Managing Troublesome Production Problems 223
Kaizen for a Smoother Flow of Goods 228

9 Standardization for Decision Making 239
Standardization of Meeting Preparation 239
Standardized Procedures for Responding to Complaints 242
Thorough Review of Plant Investment Plans 246

10 How Standardization Can Help U.S. Operations 253
Clear Presentation of Standardization Concepts 253
How to Make Standards Really Understood 255

About the Author 259
Index 261

Publisher's Message

Standardization is one of the best-kept secrets of Japanese manufacturing success — a hidden jewel for producing high quality with low cost and fast delivery. In an age of continuous improvement, six sigma, zero defects, and ISO 9000, competitive quality, cost, and delivery are not options but requirements of your customers. It is truly impossible to set — or meet — any target of excellence in quality, cost, or delivery without standardization.

We in America have used the word *standardization*, and of course have supplied and utilized its concepts in our manufacturing enterprises for at least a hundred years. Most of us, however, have not been able to truly standardize our companies and do continuous improvement at the same time. In the past, because cycle time and delivery time were long — months, or in some cases, years — we were able to handle change slowly. But today, with cycle time down to hours, minutes, or even seconds, we must be able to control these vast changes quickly and in a very simple way.

Shigehiro Nakamura is a true genius in his ability to communicate the best of Japanese standardization methodologies for the benefit of Western companies. Nakamura, who is a total productivity management consultant with the prestigious Japan Management Association, teaches that standardization is both the starting point and the culmination of continuous improvement. Starting with the current conditions, you first make a standard of the way things are done now. Then you improve it, check to see whether the improvements have the desired effect, and standardize again on a new, more effective method. Continuous improvement is the repetition of this Plan-Do-Check cycle.

Standards can be technical, procedural, or both. The large part of this book is about simplifying procedures and building-in a hierarchy for continuous improvement. Following standards can reduce costs substantially and systematically improve product and process quality. The beauty of Nakamura's approach is that it brings together methodologies like quality control, production management, industrial engineering, and total productive maintenance, which are usually scattered across an organization and used by different groups of people. Nakamura merges these methods into a unified approach you can use to produce your own road map for world class competitiveness.

In *The New Standardization: Keystone of Continuous Improvement in Manufacturing*, Nakamura outlines a new approach to standardization that goes far beyond the image of mountains of notebooks and files that simply collect dust. To Nakamura, standardization means a living system of *just-in-time information* that documents and teaches the best methods and delivers to employees the information they need, exactly when they need it, and where they need it. This kind of active standardization shares information about the best way to do things and tracks what needs to be done, ensuring high quality, low cost, and fast delivery — and total customer satisfaction.

Unlike traditional standardization methods, Nakamura's approach is dynamic and participative. It relies on the knowledge and experience of the employees who will use the standards. Nakamura emphasizes the importance of creating standards that are easy to follow — so that they *will* be followed — and can be changed as needed to reflect improved methods.

Improvement, in fact, is inseparable from standardization in Nakamura's thinking. This holds true on several levels. At a basic level, standardization is the culmination of the Plan-Do-Check improvement cycle described earlier. You analyze the operation or the system and develop a better way to do the job, check the results, then put means in place to make sure the job is always done that way, whether a veteran worker or a new hire is doing it.

From a higher perspective, a company can actually improve the way it performs standardization itself. The author visualizes workplace standardization improvement as stair-steps that represent increasing levels of standardization technology and power. The ground level is the traditional way of conveying information to individual employees through person-to-person instruction in work methods. This personal approach is subjective, however, and the teacher may not always be available to give assistance during the learning process. Documenting the instruction in the form of a manual, then, is a slightly higher form of standardization that preserves work methods for others to learn.

Manuals, however, have their own flaws. In most organization, they are dead documents sitting on a shelf, outdated and unused. The next level of standardization improvement is to make documentation for procedures and the like as visual as possible and place it right where the work is done. This can take the form of illustrated, step-by-step standard procedure sheets placed directly in the work area. The key is to put the information where it will be used — and to keep it current with improved methods developed by the employees themselves. Advanced companies, such as the Matsushita Washing Machine plant, even install work-site computers to bring complete, up-to-date production information and manufacturing instructions to employees on a just-in-time basis.

From improved ways of giving information and instructions to employees, the next higher steps in standardization involve increasing levels of automation. The simplest level is installation of poka-yoke (mistake-proofing) devices that either prevent inadvertent errors from creating defects or pull defects from the production stream. At the highest level is complete automation of operations and standardization of materials so that the work is done the same way each time, with no human error. This liberates people from machine-watching and enables them to do more creative work.

Nakamura's book examines various types of standards appropriate to the workplace, how to make them visual for workers and managers, and how to build them into machinery and equipment. In comprehensive, detailed chapters Nakamura addresses applications of standardization methodology to quality improvement, equipment management, operations improvement, and production flow scheduling.

Standardization is truly the keystone for ongoing improvement — without it, your best efforts will be temporary or incomplete. Whether you are a manager, an engineer, a technical skills trainer, or a work-site supervisor, your work will benefit from a clearer understanding of standardization. Nakamura's genius is in showing how applying a few simple principles can improve and keep on improving nearly every aspect of work life, including administrative work.

It is a privilege to be able to publish this important book in English. We would like to express our indebtedness to Shigehiro Nakamura for working with us in making his work accessible to Western readers. Thanks to Kazuya Uchiyama, Hiroshi Shimizu, and Japan Management Association (JMAM) for making it possible for us to publish the book. We appreciate as well Bruce Talbot's work on the English translation.

A number of Productivity Press employees deserve thanks for outstanding work on the preparation of this book. Sally Schwager worked extensively with the author in Japan to clarify the material; Mugi Hanao also helped answer translations questions; Karen Jones edited the text; Dorothy Lohmann and Karen Jones managed manuscript preparation, with copyediting by Susan Isenstein, editorial assistance and word processing by Laura St. Clair and Julie Zinkus, and proofreading by Julie Zinkus and Scott Hacker; David Lennon managed production, with composition by Gayle Joyce and art preparation by Gayle Joyce and Karla Tolbert; Gary Ragaglia designed a beautiful cover that reflects the dynamism of the subject matter.

Special recognition must also be given to Connie Dyer and Virginia Hallman for their extensive work to develop English materials for the author's American seminars, and to Joyce Barnes for art preparation. Their work on these related materials also greatly enhanced this book.

Norman Bodek
President, Productivity, Inc.

Foreword

American manufacturers are intrigued with the Japanese approach to continuous improvement. In trying to understand the Japanese improvement process in manufacturing, some of us make the long trip to Japan to see the factories. As visitors, we get a brief opportunity to see the process in action, to make notes, and to ask questions. After our visits, we try to adapt and implement some of what we have seen into our own company. When the implementation sputters, it is tempting for us to cite the differences between the American and Japanese cultures as the cause of the difficulty. We want to say that what works in Japanese manufacturing cannot work in American manufacturing because Americans and Japanese are different. This reasoning is hollow, though, and we feel it.

It is true that we are culturally different from the Japanese, but culture is not the reason we have difficulty in adapting and implementing successful Japanese approaches into American manufacturing. Japanese manufacturers also encounter obstacles in promoting continuous improvement. They overcome these challenges the same way we can. Unfortunately, during our short visits to Japanese factories, we see only the tangible effects of the continuous improvement effort. To a visitor, it seems that Japanese managers merely have to issue orders and the Japanese workers follow the orders faithfully. We cannot see the approaches and techniques used over the years to reach the high level of success. We don't see the continuous planning, evaluation, and revisions supporting the improvements, or the deliberate and constant effort required to implement and maintain even the simplest improvements. Standardization is the essence of all these "invisible" activities that support continuous improvement and world-class results.

Since the improvements look easy to do, we assume they are easy to implement and maintain. When we try to implement similar concepts in our American factories, however, we encounter difficulty and feel as though we have missed some key factor of success. Actually, there is no missing key. The process behind successful continuous improvement is the same in Japan and the United States. The process is so basic we tend to dismiss it. It is simple. Continuous improvement requires continuous effort. Through standardization, we assess our current condition, plan and test improvements, and stabilize at a new level of performance, over and over again. It's that simple.

Those who are really doing continuous improvement seldom have the time to explain the detailed steps of their accomplishments, so I am grateful to Mr. Shigehiro Nakamura for the time he has taken to write this book. For too long, the Japanese approach to continuous improvement has received only superficial analysis in American publishing. Mr. Nakamura's text finally gives us an in-depth look at standardization as the foundation for continuous improvement. He explains for us the invisible process we could not observe on our plant visits.

Although continuous improvement is simple, it is not easy. Fortunately, Mr. Nakamura gives us complete descriptions of approaches and techniques we can use to improve our companies. Thanks to him, we now know there is no secret to continuous improvement. We can stop our unnecessary search for a missing key. It is time to adapt what we can, roll up our sleeves, and begin the hard work. It *is* simple. Continuous improvement requires continuous effort.

Mark O'Brien
TPM Specialist
Yamaha Motor Manufacturing Corporation of America

Preface

On entering the realm of production on a manufacturing plant floor, most people are overwhelmed by the variety of themes to deal with. QCD (quality, cost, and delivery) can sum up the main themes for every factory: making good products reliably, easily, safely, cheaply, and quickly. For many years I have grappled with these themes, and when I have worked out solutions that enable the workshop or factory to make good products easily, quickly, inexpensively, safely, and reliably, it has brought me great satisfaction.

In the course of my work, I realized that the process of developing these kinds of solutions has never been standardized. This is rather strange, considering that, in general, standardization is helpful. How is it that industrial engineering techniques have been studied everywhere but standardization has never been applied to them?

Until now, standardization has been limited to the creation of standards manuals, most of which are rigidly logical in their approach, sometimes to the point where the standards are off the mark. Standardization is pointless if it fails to offer useful information to the factory employees. This book begins with a problem-conscious approach and also presents standardization techniques based on my first-hand experiences in various work sites. Promoting user-oriented standardization was the starting point for this book.

In writing this book, I wove a common thread through my experience-based explanations: All of my explanations of standardization techniques are made with the user in mind. Explanations of standardization techniques are useless unless it is clear who will benefit from them. In this instance, the beneficiaries are the people who work in factories. Standards are the foundation for the operations carried out by these workers. This book contains descriptions of approaches and techniques that factory people can use to create standards.

I wanted this book to have another common thread as a foundation for the operations of factory workers: A useful definition of standards. Wishing to avoid an abstract, theoretical definition, I reflected on my perspective and came up with the following three statements:

1. Standards are clear, simple descriptions of the best methods for making things.
2. Such standards come from improvements made to our best current approach.
3. Standards must be observed, but they are only the starting points for further improvements.

I have tried to maintain this perspective when explaining all the standardization methods and examples in this book. Consequently, this book's primary purpose is to answer three questions:

- How can standards be most clearly and simply explained to factory workers?
- How can factory workers make improvements to bring current conditions up to expected standards?
- What methods and techniques can be used to raise standards to a higher level?

We have firmly entered the era of factory automation (FA) and computer integrated manufacturing (CIM). In this highly automated and mechanized era, fewer people are working in the factory. Does this shrinkage of the human role make standards for human work less or more important? I argue that the higher the standards for human work, the easier the transition to automation; higher standards enable operators and managers to understand their roles more clearly.

I place an especially strong emphasis on improving standards. Standards should not last forever. The idea that standards should not or cannot be improved is pessimistic. Standardization should be optimistic: There is always room for improvement and a better future.

Today's factories are changing rapidly under the impact of technological advances. People who work in factories can be certain that their relationship with the production equipment and machines (including computers) will continue to change. But no matter how this change occurs, the human activity of standardization will remain important. This book takes this into account.

Having a clear idea on how to improve quality, shorten lead time, and reduce cost is indispensable to improving the production system. A number of methods, such as industrial engineering (IE), quality control (QC), and just-in-time production (JIT), address these issues. My 20 years of experience with these methods, however, has shown me that applying these methods in isolation of each other has little effect. Collecting and analyzing statistics out of the context of the workplace yielded neither fast improvement nor fruitful effects. I also realized, through working on improvements in the workplace, the importance of fact finding and pursuing real solutions to root causes.

I have found that it is more effective to integrate these methods. Organizing the improvement program according to the 5Ms of manufacturing — men/women, materials, methods, machines, and measurements — and creating a logical sequence for their implementation are essential for achieving high-level, steady, safe, and easy improvements. Incorporating a visual control system into these methods and involving all employees are also key elements for improvement.

This integrated approach is discussed throughout the book as a basic standardization technology; for ease of reference, I call this technique the "SN" method or approach. This method has generic characteristics and has been applied in various industries, as the examples show. It addresses standardization issues in several facets of manufacturing operation, including quality, operation methods, equipment availability rates, information systems, and decision-making processes (each is discussed in a separate chapter of the book). Through two and a half years of experience as a line manager in a Japanese-owned company in the United States, I can also say that the SN method is a universal approach that can help overcome cultural differences among employees and reinforce standardization in the workplace.

In closing, I would like to offer my heartfelt thanks to the people I have worked with in many work sites at my former employer, Hitachi Metals, and at various affiliated companies, all of whom were very helpful in supporting my research.

Shigehiro Nakamura

1
Standardization for a New Era

ERRORS IN THE CONVENTIONAL APPROACH TO STANDARDIZATION

In factory operations, errors in standardization fall into two main categories:

1. Errors in describing methods and procedural sequences (missing words, for example).
2. The use of standardization approaches that are unsuitable for the particular user or situation.

The second category of error is more common, primarily because of a lack of careful attention. Suppose, for example, that your approach for reducing defects is to have equipment operators carefully go over each major and minor item on a long checklist. This may require a lot of worker training, especially for newer employees.

Suppose that the checklist consists of 3 items the operators can check off at a glance, 6 to 12 items that require some decision making or judgment call, and 30 more items that the operators added on their own.

While using an extensive checklist may seem logical, as a practical matter a long checklist is too difficult to use effectively. This is a typical example of a theoretically correct approach to standardization that is inappropriate for the work site. Figure 1-1 lists three typical examples of mistaken approaches to the application of standards.

Mistaken Approach	Resulting Problems
1. Posting (or teaching) a detailed list of standards at the worksite in attempt to eliminate problems.	The list is too long and difficult to complete, so few of the standards are observed (over-standardization).
2. Worksite instructions are neatly and clearly printed and displayed, but some points have been changed or are otherwise irrelevant.	The list does not suit the current needs and tasks of the targeted workers (inappropriate standardization).
3. A system of meticulous management to minimize problems has been in effect for a long time, with little progress. Detailed records are kept to support standardization, but the workers are too concerned with record-keeping to question current practices or look for improvements; improvements have been limited to adding new items to check when new problems occur. The overriding concern has been to maintain the status quo.	No steps have been added for eliminating QCD waste. This approach does not promote progress (overconservative approach).

Q = Quality, C = Cost, and D = Delivery

Figure 1-1. Mistaken Approaches to Standardization

Standards are important tools for improving companies. To make standards effective, you need to learn to use them systematically and appropriately. Figure 1-2 describes the correct approach to creating new standards.

Standardization starts with ideas or concepts. The ideas presented in this book will help you take a fresh look at standardization. The examples describe the approaches and methods for putting these ideas into practice. The key concepts of standardization include:

1. *User orientation.* Standards should be made expressly for the people who will use them and should help these employees do their work more reliably, easily, safely, and quickly.
2. *QCD integration.* Quality, cost, and delivery (QCD) are key factors that should be integrated into and accounted for in each standard. Stan-

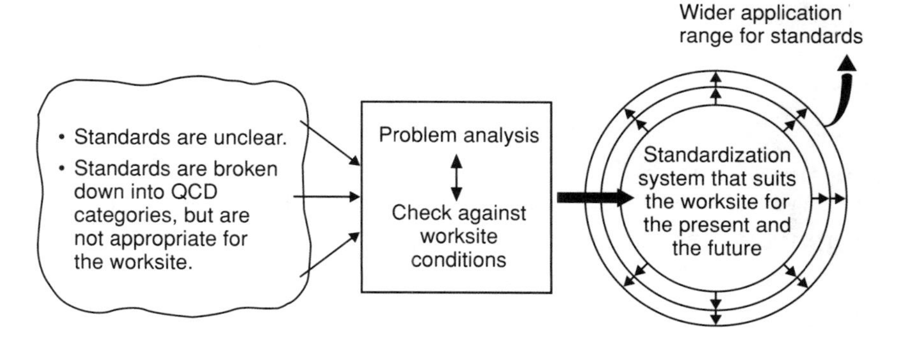

Figure 1-2. Approach to New Standards

dardization in which control points or checklist items are simply added as needed should be reviewed and improved to integrate QCD. The most effective management comes from using a few control points that are presented clearly.

3. *Standards improvement.* Standards must always be improved to keep pace with the latest technological advances. The people who make and keep the standards must maintain an improvement-minded perspective. It bears repeating that standards are tools that should help people make good products reliably, easily, cheaply, and quickly. Tools are always subject to improvement. When using the latest revised standards, stay open to the possibility of further refinements.

Keep in mind these three concepts: they are basic to finding the right approach and checking results in making and enforcing standards.

Another basic "must" for standardization is problem consciousness. The following list describes problems frequently encountered in standardization.

1. The job performances of new hires and veterans are vastly different. Standards manuals generally neglect to say how newcomers can become veterans.
2. Standards manuals usually contain many check items, and the significance of the check items is rarely noted.
3. Standards manuals often fail to describe the course of action for situations in which measured values are outside the control levels. The control points may be described well enough, but workers have to look elsewhere to find out how to respond.

4. Standards manuals often go unrevised for as long as three years. After a while workers are no longer aware of the purpose of certain check items.
5. Recording actual values is a waste of time when they are normal.
6. Often, the check items have not been prioritized. Workers can easily overlook the more important items since they do not know which are more vital than others.
7. Standards manuals sometimes contain obsolete checking procedures that automated monitoring and data recording devices now handle.
8. Standards manuals may call for data recording that is now unnecessary due to the use of *poka-yoke*, or mistake-proofing, devices.* Standards manuals are sometimes written even though they have been made unnecessary by safety measures.
9. There may be two separate standards manuals, one for equipment management and another for quality control, that discuss virtually identical points.
10. Sometimes much time and energy go into producing neatly printed standards manuals that fail to use the standard procedures documented in workers' notebooks.
11. The layout of standards manuals may make them difficult to read and implement. The sequence of steps and the collection of standards into a system should suit the perspective, education, and work site conditions of those who use the manuals.
12. A well-written, detailed standards manual is sometimes worthless because the checking or inspecting of a previous process had already covered all the points in the manual. Try to avoid unnecessary double-checking.
13. Often check points are added to standards manuals to address problems whose causes were never fully investigated.

---

* *Poka-yoke* (mistake-proofing) devices have a range of types and applications. Many such devices prevent measurement or adjustment error by using block gauges and the like to avoid measuring and data checking altogether. For more on the subject, see Nikkan Kogyo Shimbun, Ltd. and Factory Magazine, eds., *Poka- Yoke: Improving Product Quality by Preventing Defects* (Cambridge, Mass.: Productivity Press, 198); Shigeo Shingo, *Zero Quality Control: Source Inspection and the Poka-yoke System* (Cambridge, Mass.: Productivity Press, 1986). — Ed.

14. Some standards manuals contain explanations that have either technical terms or oversimplified explanations. Both situations fail to give the user adequate understanding. A user not involved in the writing should review the standards manual.
15. Some standards manuals have no systems for review or evaluation. Conversely, other manuals are reevaluated and revised so often that they become too complicated.

Improvements start with problem consciousness. Take some time to reevaluate the written and physical standards (standards sheets, displays, tools, jigs, etc.) at your company in light of this list of common problems.

The following sections examine different measures that can be taken for various work site conditions.

STANDARDS INDICATE THE CURRENT LEVEL OF PRODUCTION TECHNOLOGY

Since standards are an important tool for improving the overall quality of a company, companies need to understand how the development of standards fits into the company's inner workings. One way is to ask "what if no standards existed or the standards were too lenient?"

What If Quality Standards Are Nonexistent or Too Lenient?

When standards for building in quality are missing or inadequate, many problems can arise (see Figure 1-3). If there are no standards for supplying materials (materials and parts), defective parts will be overlooked; when fed into the production process, these parts produce a defective product, which creates waste. Likewise, if there are no standards for the machines, operation methods, and measurement methods used in the production process, process operators' work becomes sloppy and product quality varies greatly. The result is defective products and customer complaints that eventually destroy the product's sales strength.

Obviously, all manufacturers realize this and have established rules for the design of production processes, the use of tools (equipment, jigs, materials, etc.), and the sequence of operations. Companies are obliged to set standards that answer the "5W1H" questions: When, Where, Who, What, Why, and How.

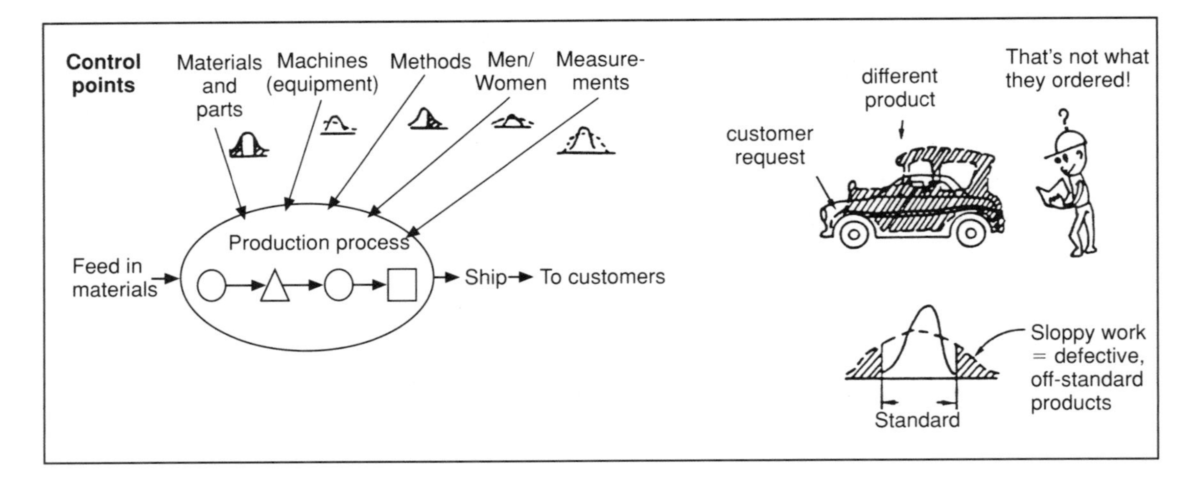

Figure 1-3. Elements for Building Quality into Products

Recognizing such standards as a type of production technology is important. It is a technology that is essential for determining the quality of products, the cost of producing them, and the degree of excellence in the QCD factors. Figure 1-4 describes various quality standards related to the product development cycle.

Although some standards come from outside the company, such as organizational or national regulations and client specifications, most tools for building quality into production come from in-house standards. When a company can maintain a high level of in-house standards, it is better able to make products that satisfy customer needs, as shown in Figure 1-5.

What If Operational Standards Are Nonexistent or Too Lenient?

If no standards existed for operating methods or machining times, it would be impossible to determine the occurrence of production defects, to plan production schedules, or to manage production in general. Figure 1-6 illustrates the importance of standards to production management.

Under the "check" circle in Figure 1-6, are a picture of a balance and the words, "If the target is not correct, control of results is impossible." What does *control* mean in this case? Consider the following formula as a simple definition:

$$1 \begin{smallmatrix} < \\ = \\ > \end{smallmatrix} \frac{\text{Results}}{\text{Target}}$$

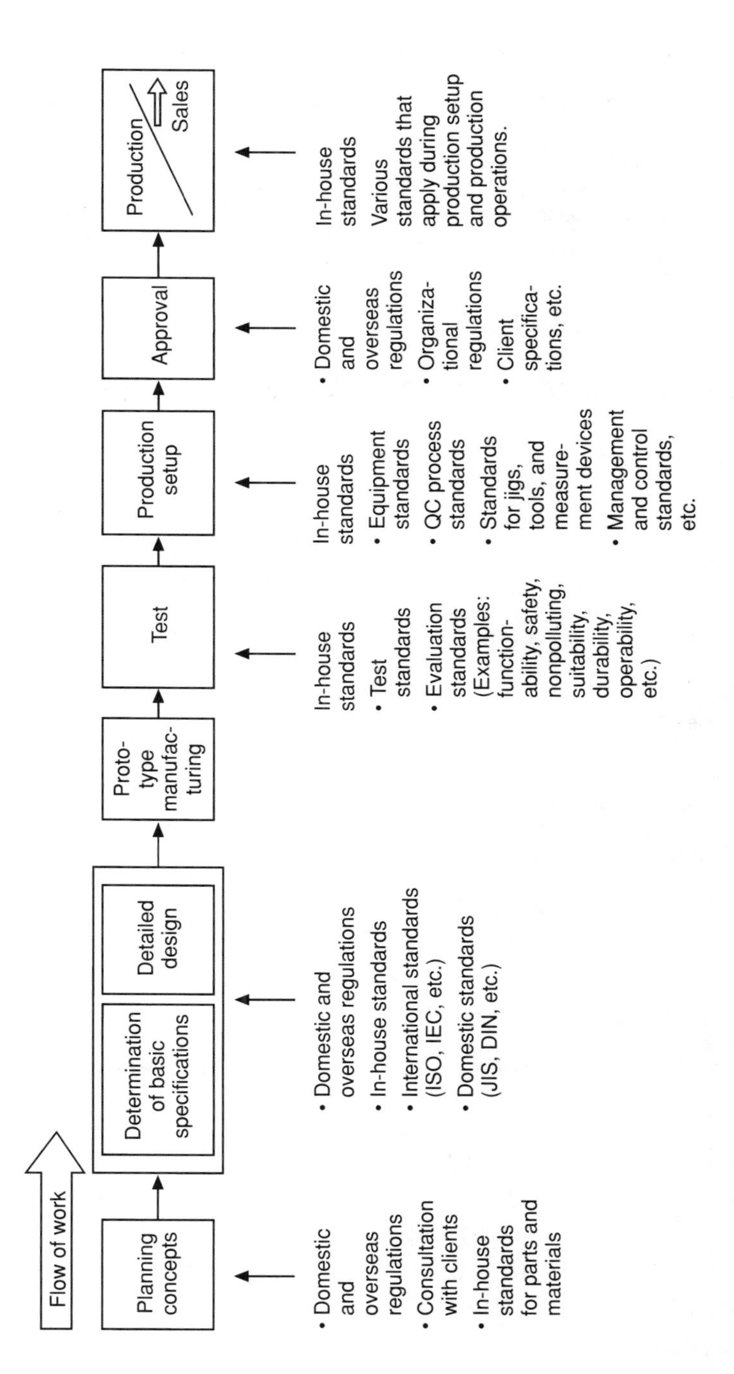

Figure 1-4. Flow of Work and Types of Quality Standards

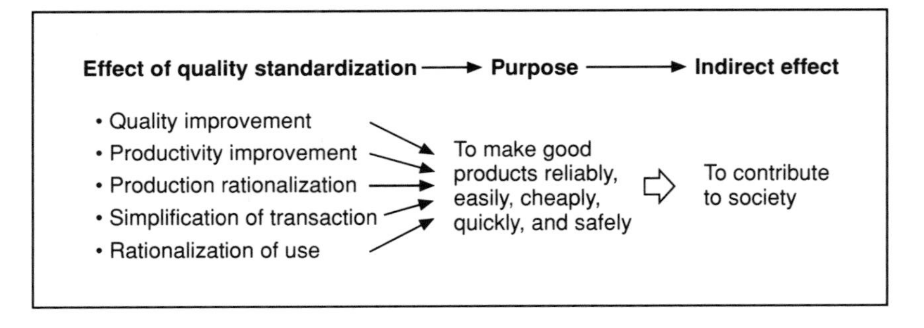

Figure 1-5. Purpose and Effect of Quality Standardization

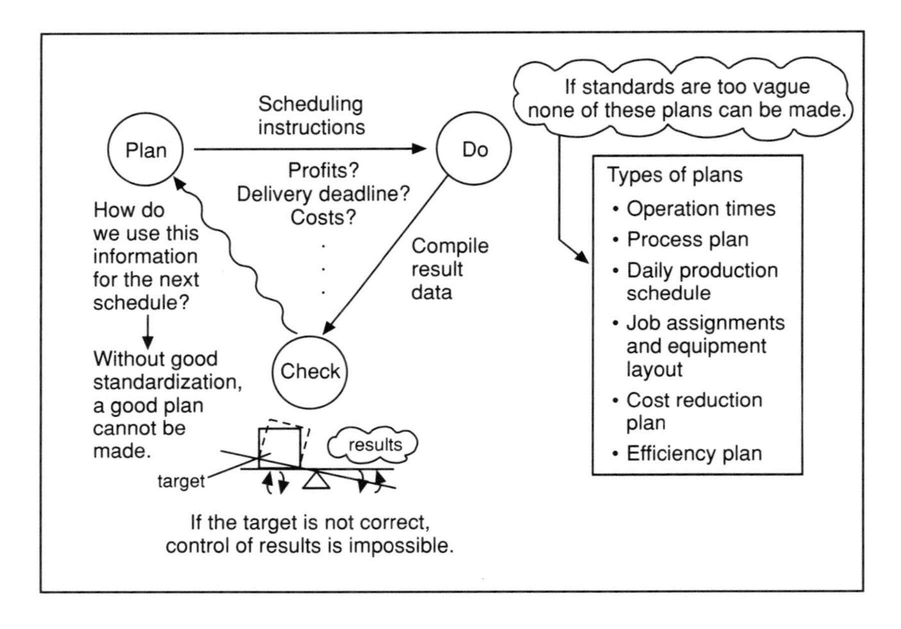

Figure 1-6. The Importance of Production Time Standards

This means that control is a comparison of result values against target values. The closer they are, the closer they come to the value of "1" on the left side of the equation. If the result values do not reach the target values, there is a problem somewhere, usually having to do with waste of some sort. When the process or factory is able to keep the value of results/target at 1.0 or more, it is producing nondefective products. Conversely, if the value of results/target falls below 1.0, there is a quality problem.

You can use this formula as a simple criterion for judging production defects. When you find it easy to reach the target value, try to set a higher target to work toward. This is what improvement activities are all about.

In sports, track athletes train to achieve a competitive race time against which they compete. When their performance drops below that time, they find out why and train to overcome the problem. When they reach the target time, they set themselves more difficult goals to work toward. Athletes and companies improve their performance in similar ways — by repeatedly working toward higher goals.

Figure 1-7 shows a diagram developed by Shigeru Ono to illustrate how standardization promotes the health of companies. If no standards existed, companies would waste their energy and resources in dealing with problems and would have little left for preventive analyses and improvement activities. The bracketed paragraphs at the bottom of the figure deserve special attention. This figure underscores the importance of standardization as part of any company's activities.

STANDARDIZATION FOR THE FA/CIM ERA

The pursuit of standardization is a key theme for the current era. Figure 1-8 summarizes several historical stages of development in the manufacturing industry, showing the steady procession toward standardization and automation.

In the early nineteenth century, standardization and mechanization enabled the division of labor in production operations. Products were standardized using the principle of interchangeable parts. In the later nineteenth and early twentieth centuries, the research of Frederick W. Taylor and others laid the foundation for industrial engineering and began the process of standardization for operation times and methods. This phase of standardization led to improvements in worker productivity.

By the middle of the twentieth century, organizations such as the International Standards Organization (ISO) were formed to help set standards that promote progress in mass production, for such products as automobiles and their components. Today standardization continues to play an important role in promoting wide-variety, small-lot production that is more responsive to fast-changing markets. New standards enable companies to respond effectively to market changes with flexible and efficient production.

Some of the ideas about standardization presented in this book date back to Henry Ford, pioneer of the conveyor assembly system and various other forms of standardization.

> The eventuality of industry is not a standardized, automatic world in which people will not need brains. The eventuality is a world in which people

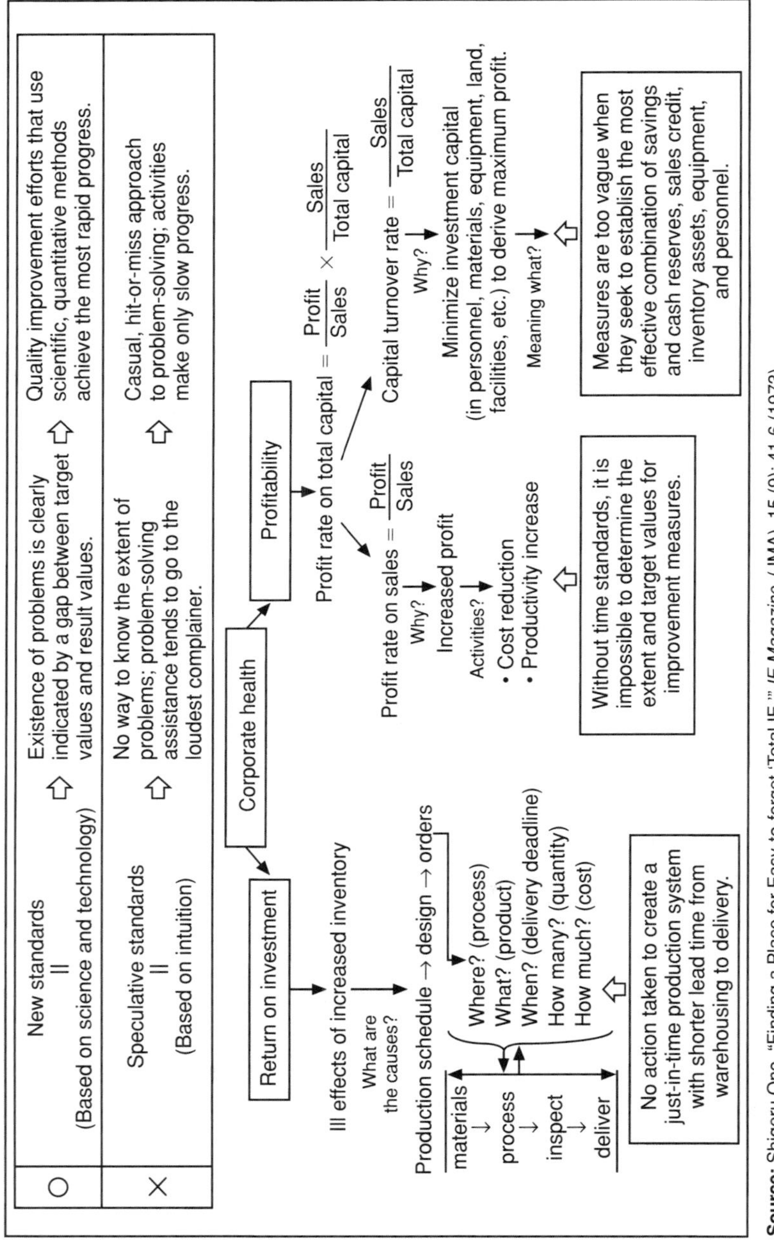

Source: Shigeru Ono, "Finding a Place for Easy-to-forget 'Total IE,'" *IE Magazine* (JMA), 15 (9): 41-6 (1973).

Figure 1-7. Standardization: A Foundation of Corporate Health

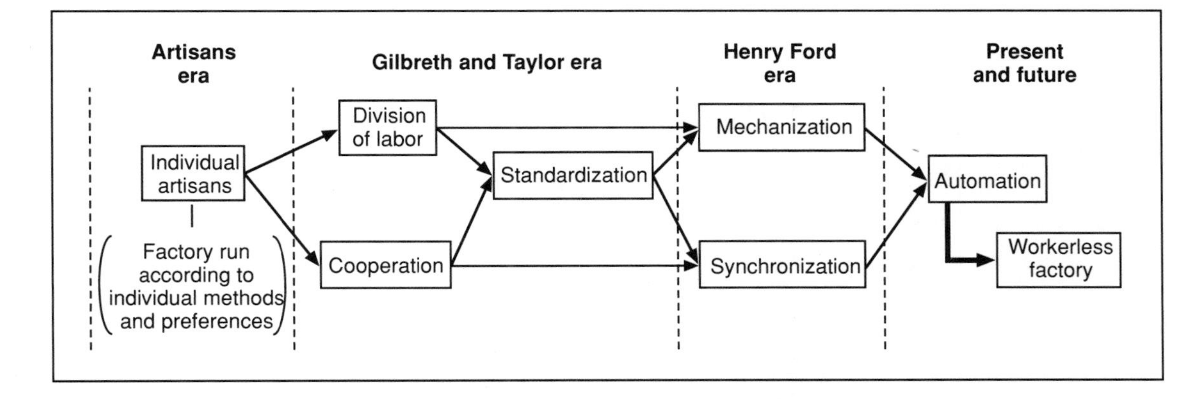

Figure 1-8. History of Manufacturing: Toward Automation

> will have a chance to use their brains, for they will not be occupied from early morning until late at night with the business of gaining a livelihood. The true end of industry is not the bringing of people into one mould [sic]; . . . industry exists to serve the public of which the working man [sic] is a part. The true end of industry is to liberate mind and body from the drudgery of existence by filling the world with well-made, low-priced products. How far these products may be standardized is a question, not for the state, but for the individual manager. — Henry Ford, *Today and Tomorrow* (New York: Doubleday, 1926; Cambridge, Mass.: Productivity Press, 1988).

We live and work in an era of diversification and constant change. One thing that has not changed, however, is the need for manufacturing companies to raise efficiency by making the parts, operations, and equipment they use as uniform and standardized as possible. Certain applications of computer technology have enabled companies to respond more flexibly to market changes. Figure 1-9 shows how standardization can help companies meet the needs of this new era.

As the figure shows, standardization techniques are far from obsolete and in fact can serve as a fundamental support for companies striving to keep up with the times. In particular, standardization that aims to integrate the essential QCD factors, such as that which builds quality into the production equipment, is especially important for today's factory automation and computer-integrated manufacturing.

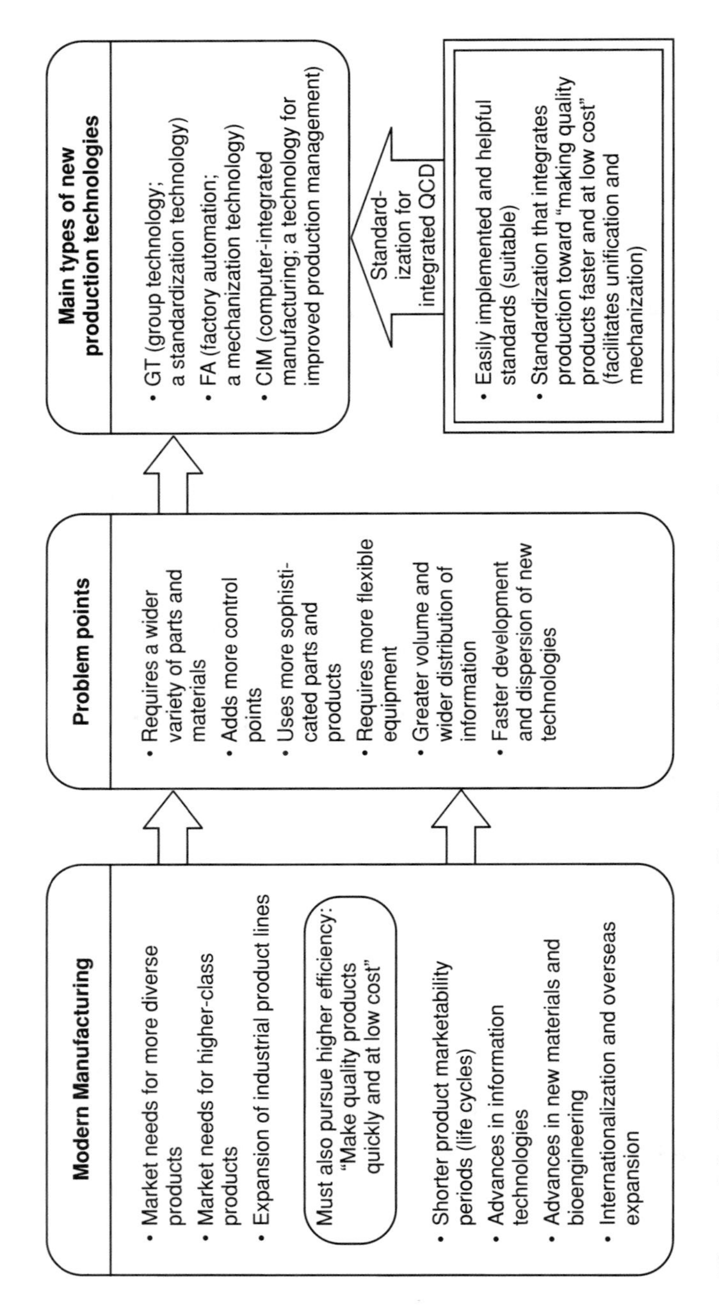

Figure 1-9. How Standardization Can Respond to Today's Manufacturing Needs

STANDARDS ARE MADE TO BE IMPROVED

What specific kinds of standards are needed in this new era? To understand this clearly, we will first define what we mean by standards, then look at how they are categorized into operation-based categories and improved.

What Are Standards?

Let's look at some definitions of standards that have been proposed by various sources:

- A standard is a rule or example that commands respect and that is established by a recognized authority based on custom or consensus.
- Standards are based on the summary results of science, engineering, and experience.
- Standards are established to facilitate current as well as future progress.
- Standards are technical specifications that are significant and should be followed.

The following characteristics of standards can be drawn from these statements:

1. Standards provide clear indications.
2. Standards are scientific.
3. Standards should be respected and followed.

Note that none of these characteristics provides a specific context. Are we describing standards in the sense of regulations? Are they organization standards such as the JIS series developed by the Japanese Standards Association? Or are they specific descriptions of operation methods? For the purposes of this book, we will deal with standards used in the company and factory context; therefore we will define the term as follows:

> Standards: Written and graphic descriptions that help us understand a factory's most trusted and authoritative techniques and provide knowledge about various production-related themes (men/women, machines, materials, methods, measurement, and information), with the aim of making quality products reliably, easily, safely, inexpensively, and quickly.

The goal of "making quality products reliably, easily, safely, inexpensively, and quickly" is included to emphasize the goal-oriented nature of standardization. The "techniques and knowledge" aspect of the definition refers to the criteria and procedures for making things. These criteria and procedures are practiced and honed until they become authoritative within the company. The part about "written and graphic descriptions that facilitate understanding" underscores that standards should be clear and easy to understand to ensure that they will be used correctly. In addition to textual information and graphics, the specified jigs, tools, measuring devices, and other equipment can also be viewed as "standards" for factory operations. This aspect of standardization will be described later.

Categories of Standards

Standards can be categorized in three ways: basic characteristics, application, and the object of standardization.

Standards Categorized According to Basic Characteristics

Basic characteristics of standards include range of jurisdiction, type of contents, level of enforcement, and period of validity, as shown in Figure 1-10. Technical and process standards — two basic categories of corporate in-house standards — are described in Figure 1-11.

(Boxed items are discussed in the text.)

Application or Jurisdiction	Type of Contents	Level of Enforcement
• Corporate in-house standards • Organizational standards • National standards • Regional standards • International standards (Organizational, National, Regional, International: Public standards)	• Basic standards (terminological standards, test method standards, etc.) • Quality standards (product quality, procedural standards, service standards) • Operation standards and equipment maintenance standards • Standards labeled by purpose (interface standards, compatibility standards, variety reduction standards, safety standards, product protection standards, etc.)	• Technology standards (including legal standards) • Implementation standards (including legal standards) • Mandatory standards • Elective standards • Temporary standards • Provisional standards

Source: *In-house Standardization Handbook*, 2d ed. (Tokyo: Japanese Standards Association, 1989).

Figure 1-10. Categorization of Standards

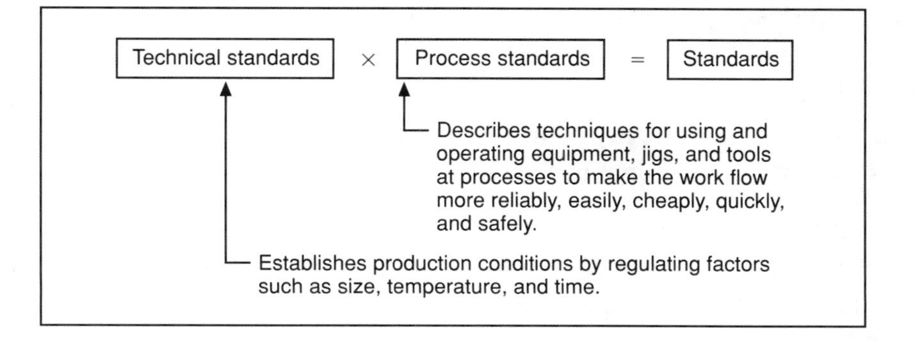

Figure 1-11. Formula for Establishing Standards

Standards Categorized According to Application

Since standards are used in various workplace situations, categorizing them according to the cases in which they apply makes sense. We can establish such categories by defining common workplace situations and determining how the methods specified in the standards can best serve the objectives of workplace operations. Figure 1-12 lists some specific examples of standards categorized by workplace application.

Application	Examples
1. Standards with technically complex material that must be understood well before use.	Classroom situations; textbooks, standards manuals, etc.
2. Standards used during production in checking measured values against technical standard values.	Ordinary technical standards, check sheets, etc.
3. Standards used to monitor abnormalities or search for causes in response to alarms.	Standards used in poka-yoke measures, observation, and investigation.
4. Standards that describe the sequence of manual operations.	Standards used in standard time systems.
5. Standards that check progress (operation records, output totals, quality results) to discover problems and indicate where improvement activities are needed.	Standards whose purpose is to collect data that will uncover waste in the production system.

Note: This application-based categorization is also useful for clerical operation standards, in which case the standards concern the production and flow of "information" instead of "things."

Figure 1-12. Standards Categorized by Application

Standards Categorized According to Object

Object-based categorization of standards enables you to see more clearly which standard is most appropriate for each situation. Reasons for categorizing standards according to object include the following:

- Clarifying the purposes for which the standard is used.
- Identifying standards that may serve a different purpose but can still be used for the object in question.
- Identifying standard applications for different stages and levels of production, which helps you increase the effectiveness of standardization and more smoothly apply standards to these object categories.

Figure 1-13 shows how standards can be categorized according to object.

What Is Standards Improvement?

Generally, we think of standardization as the task of putting together standards manuals. Today, standardization includes much more than that. It also involves improving the methods, with the ultimate goal of making compliance something that anyone can do automatically. For example, written dimensions in

Object	Examples of application-based standards
1. Things	Quality control standards, inspection standards, standards for responding to customer complaints
2. People and operations	Standards describing the sequence of manual operations, work instruction standards, etc.
3. Equipment	Equipment maintenance standards
4. Information	Production orders, follow-up standards, item labeling standards
5. Money	Equipment investment evaluation criteria, standards for transactions with supplier companies, procurement standards, cost estimation standards, etc.
6. People and safety	Safety standards

Figure 1-13. Standards Categorized by Object

a standard for using jigs and tools to measure or control the dimensions of parts may be needless. In many cases, the objects themselves represent the standards.

Poka-yoke (mistake-proofing) devices are a case in point. A poka-yoke device, such as a positioning pin, might prevent part A from being matched incorrectly with part B instead of part A'. For a familiar example, at self-serve gas stations that sell both leaded and unleaded gasoline, the pump nozzles for the leaded gas will not fit the gas tank openings of cars that use unleaded gas. This is an example of a mistake-proofing system using built-in dimensional standards.

Building standards into objects is one level above writing standards into manuals. Figure 1-14 illustrates different levels of standards and the step-by-step path of standard improvement activities. Improvement of standards will be taken up in more detail in Chapter 3.

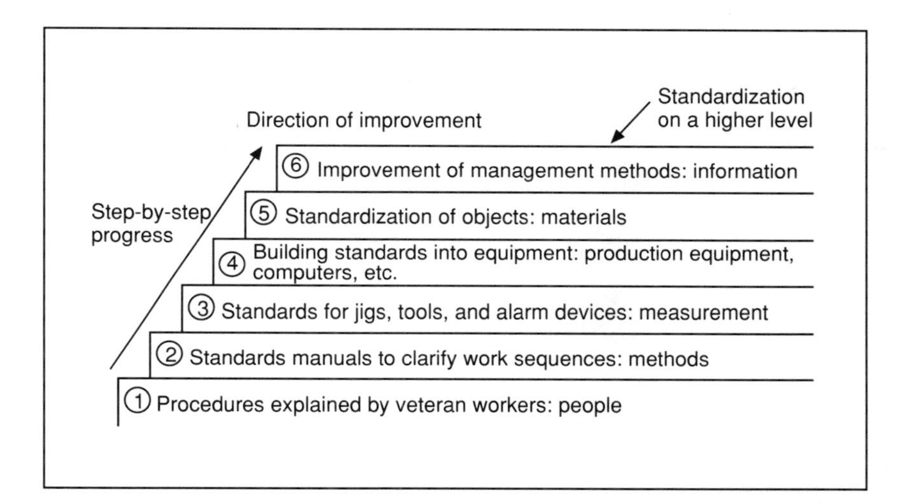

Figure 1-14. Development of Standards Improvement Activities

2
Making Quality Products Easily, Inexpensively, and Quickly

STANDARDS FROM A MANUFACTURING PERSPECTIVE

Before getting into a detailed description of standardization methods (Chapter 3), it is important first to understand how standards fit into the daily activities of a company. Figure 2-1 outlines a manufacturing company's main activities, such as receiving product orders, supplying materials to the production line, and manufacturing products. These activities can be summarized as the company's "work system."

What kinds of standards are needed to manage a work system? According to the work system outlined in Figure 2-1, materials are supplied to the system and worked on by five elements — men/women, materials, machines, methods, measurement, and information — to make products. These elements (5Ms plus one I) are the key supports for any work system that makes quality products easily, inexpensively, and quickly.

The methods used in this work system are subject to constant change. The company should continually seek ways to improve its methods. When they have developed a new method, the standards for this new method are the basis for a new round of the Plan-Do-Check management cycle. Improving methods and continually repeating the management cycle help eliminate waste and promote better production activities.

To explain further, you can be sure to find a strong work system at any manufacturing company that grounds its activities firmly in the QCD themes — *quality* product, low *cost*, and short *delivery* lead times. Such a work system requires good management in planning (plan), operating (do), and evaluating

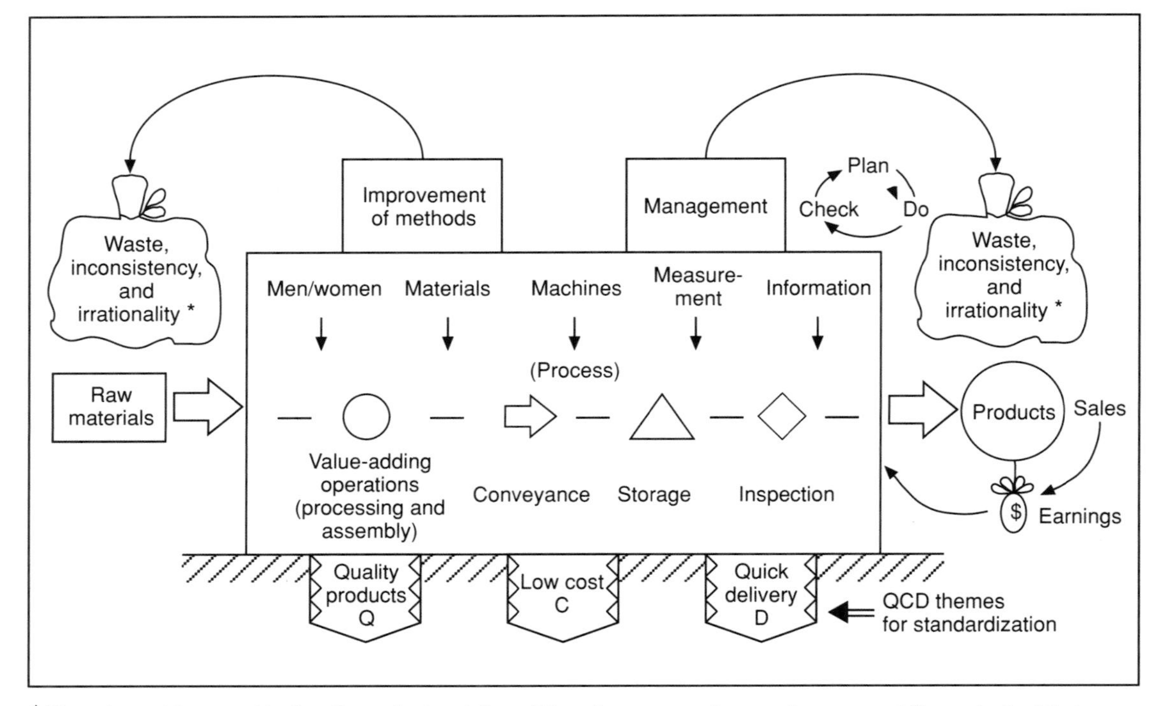

* Waste, inconsistency, and irrationality are the translations of three Japanese words — *muda, mura, muri*. These checkpoints, known as "3 Mu," refer to situations when the effort outweighs the goal, when the effort fails to reach the goal, or when results swing inconsistently between these two poles.

Figure 2-1. Elements of the Work System

(check or see). This involves working with the key elements of people, materials, and equipment. To do this, the managers must possess the technology, knowledge, and rules that explain how the elements are used. These rules are what "standards" refer to.

Management is often categorized by various production control keys or management items on the shop floor. There are five such management items, with specific standards organized under them (see Figure 2-2). Figure 2-3 outlines the types of standards directly related to production management systems.

In Chapter 1, Figure 1-4 shows how quality standards fit into the overall flow of production from product planning to sales. Figure 2-3 illustrates the kinds of standards used in the series of operations that take place each day in the production flow, from receiving orders through manufacturing and shipping.

This series of operations begins when orders come in from customers. From this point, information must flow to each process to describe the type of

Management item	Standards (examples)
1. Management items required for production activities	• Acquisition and storage of materials and parts • Production technology standards (in-house and external standards) for making things • Testing and inspection standards • Control standards for initial production run • Equipment installation and test operation standards • Safety standards, etc.
2. Management items required for production planning and setup	• Standards for equipment specifications, capacity guidelines, and preparation of jigs and tools • Labor-hour standards • Process patterns • Load distribution standards for planning daily production schedules • Production planning standards • Standards for work-site display boards
3. Management items for manufacturing	• Standards for managing the progress of processes • Quality control and quality assurance standards • Cost management standards • Equipment management standards (technical planning standards, installation planning standards, maintenance and PM standards) • Standards for safety measures • Standards for managing transactions with outside suppliers • Fuel and energy management standards
4. Support to help raise the level of motivation	• Suggestion system • Personnel management • Improvement in attitude and problem consciousness • Training system • Safety and environmental pollution management
5. Technical management for improved quality and higher productivity	• Human automation • Robotics • CIM standards • Standards for introducing FMS/FA, MRP, IE, QC, PM, etc.

FMS: flexible manufacturing system
FA: factory automation
MRP: material requirements planning

Source: *In-house Standardization Handbook,* 2d ed. (Tokyo: Japanese Standards Association, 1989); Mitsuru Toki, et al., *Easy-to-understand In-house Standardization* (Tokyo: Japanese Standards Association, 1987).

Figure 2-2. Factory Management Items (Production Control Keys) and Corresponding Standards and Approaches

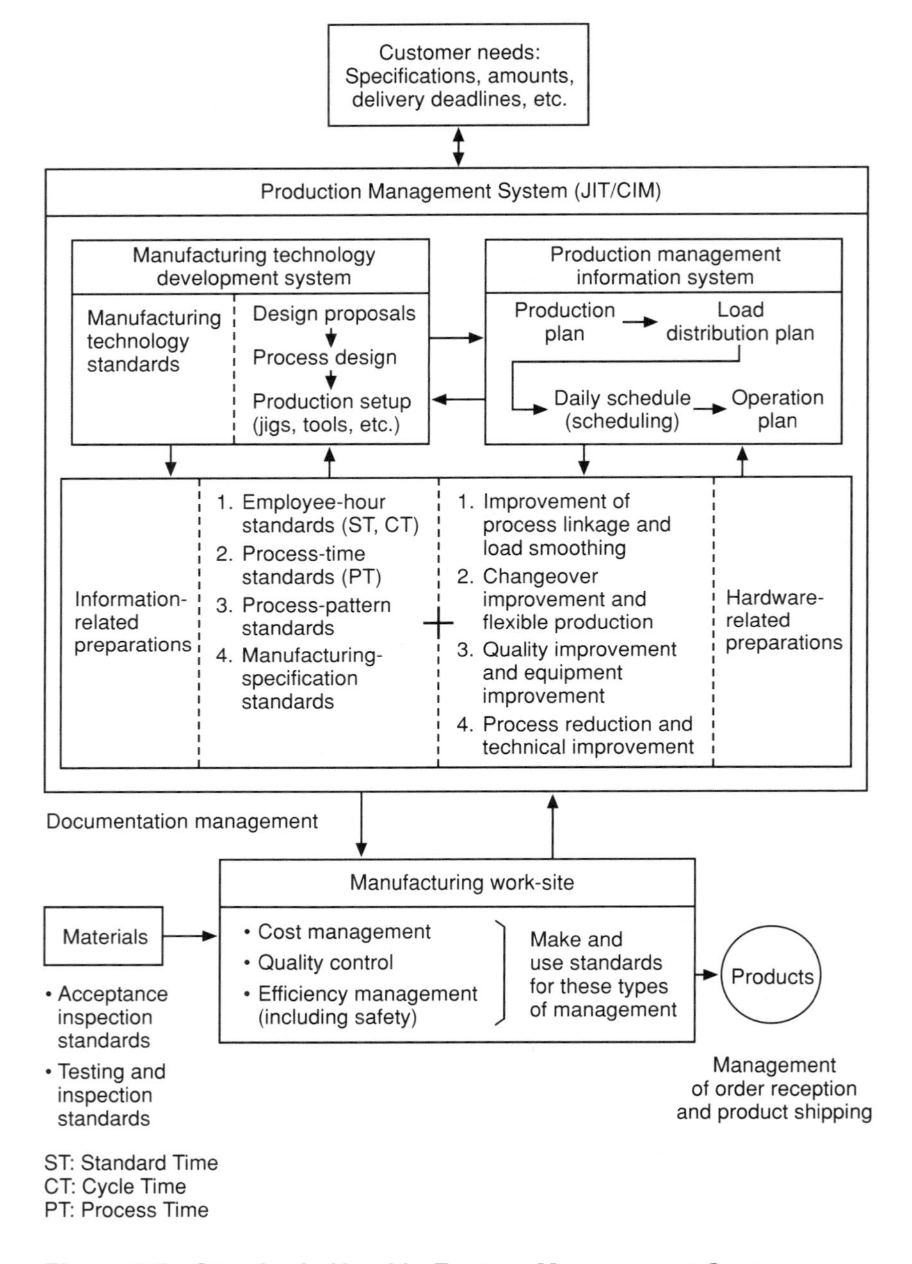

Figure 2-3. Standards Used in Factory Management Systems

product to be made, the desired quantity, the delivery destination, and the manufacturing methods used. To accomplish this, order information must be transferred between the manufacturing technology development system that makes the actual products and the manufacturing information system that relays the production information to each production process.

Two types of standards — information related ("soft") and equipment related ("hard") such as equipment restrictions and quality standards — are added to the order information as the production plan is developed. The production needs indicated by this information are then compared with various target values for cost, quality, and efficiency to prepare the final production orders. The above method is a typical production management method.

Japan's steel-making industry is currently ranked number one in the world in terms of output, quality, and productivity. Just 30 years ago, this industry ranked at a very low level in all of these areas. At that time, Japanese steel companies formed study groups that visited the world's top steel mills in North America and Europe. These study groups brought back not only hardware-related technologies but also techniques for managing equipment, people, and materials; both types of information were considered of equally high value. Soon, Japanese steel companies were cultivating their own IE and QC experts, who shared their expertise with each other. The long-term result has been Japan's ascension to the top rank in the global steel-making industry.

No matter how good a factory's equipment is, the factory cannot make good products without the help of people who have mastered the techniques for using the equipment and managing the factory. I do not categorically agree with the common assertion that Japan has the world's best production technologies. I have seen many factories in Japan that have the latest factory automation (FA) equipment but lack the expertise to use the equipment effectively. Overlooking this essential ingredient is easy, since it is not visible and obvious like the equipment.

We can use standardization techniques to develop effective ways of using production equipment and managing factories. Figure 2-3 outlines the invisible but essential role of standardization in production management systems.

STANDARDS HELP SOLVE WORK-SITE PROBLEMS

Having understood the importance of standards, it would still be incorrect to say that simply establishing standards ensures smooth production. In fact, standards or no standards, production activities are always fraught with problems and waste. When production activities do not run smoothly, the people concerned become aware that some kind of problem exists. What constitutes a "problem" in this context and how do problems relate to standardization? The following sections discuss these concerns and how to approach problems at the work site.

Why Do Problems Occur?

First, you must clearly state what you mean by "problems." Another simple formula gives a concise definition:

Ideals and Standards − Actual Conditions = Problems

In other words, when there are differences between the standards and the actual conditions, these differences are the problems.

Figure 2-4 shows two patterns for the occurrence of problems. The first pattern (Case 1) occurs when conditions suddenly deteriorate from their previous level. The second pattern (Case 2) happens when a higher standard or target has been set, and the problem lies in closing the gap to reach that higher level.

The Problem-Solving Sequence

Some problems are rather vague and murky while others are fairly straightforward and pragmatic. The latter are easier to work with because they are more amenable to the 5W1H method (the method of asking what, when, where, why, who, and how) of identifying the facts about the problem and its causes. Knowing the cause of a problem makes it much easier to solve.

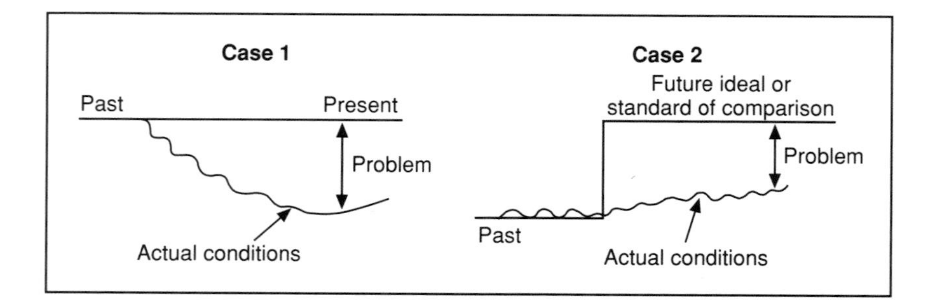

Figure 2-4. How Problems Occur

Let us look at an example of a problem-solving sequence. This example uses one of the problems from the list in Chapter 1.

Recently hired workers are turning out many more defective products than the veteran workers. Apparently the new workers still need more training.

↓

The problem is rather vague, so we visit the factory to better understand the problem.

↓

For example, at a parts machining process Worker A (a veteran worker) had recently transferred to another part of the factory and Worker B (a new hire) took Worker A's place just three days ago. The defects tend to occur at the point in the process when bushings are tapped onto the product; the defective products have damaged bush pins. The defect rate had been zero when the veteran worker was handling the process, but with Worker B, random defects have been running a very high 5 percent.

↓

What are the possible causes of this problem? The new worker's inexperience could be causing the bushing defects, which has resulted in dimensional variance during processing of the parts. Or the production equipment may have idiosyncrasies that take getting used to. Such conjecture is not likely to reveal the true cause, however. We decide to videotape the way Workers A and B handle the process.

↓

The videotape recording reveals that once the bush pin is placed on its corresponding part, it often shifts upward due to vibration. Worker A (the veteran worker) knows this and checks the bush pin each time. Worker B does not know about this tendency and has been too busy to notice it.

For the time being, we solve the problem by teaching Worker B to check the bush pin as Worker A had done. Meanwhile, we analyze Worker B's work sequence and make revisions (improvements) to facilitate holding down the bush pin.

↓

These measures reduce the defect rate, but Worker B finds it tiring to pay attention to the bush pin during the busy work sequence.

We go back to the root cause by making an equipment improvement to eliminate the vibration that had caused the bush pin to shift out of position.

↓

This reduces the defect rate to zero.

In this example, we established standard operations that provided a temporary solution to the problem; meanwhile, we worked on identifying the root cause, which turned out to be an equipment problem. We then took steps to eliminate the equipment problem, which in turn eliminated the need for the extra control point (checking the bush pin's position) that had served as a temporary solution. Eliminating the root cause is the key to making a permanent improvement.

In this case, the real problem was the effect of the equipment's vibration on the bush pin, not a gap in the technical expertise of the two workers. Unfortunately, problems are not always this simple, and the reader is therefore advised to follow the problem-solving sequence of standardization shown in Figure 2-5.

The sequence in Figure 2-5 is both a problem-solving sequence and a standardization sequence. The following subsection describes the more difficult parts of this sequence. Please refer to this section as you attempt to work through the sequence yourself.

Key points of the problem-solving sequence (and related training methods)

1. *Approach the situation with a problem-conscious attitude and thoroughly identify the problems and issues.*

Suppose you are going to travel to a tropical island for a vacation. To help you pack, you jot down a list of items you will need to take:

- shirts
- shorts
- underwear
- shoes
- toiletries

With a general list like this, however, you could still arrive unprepared. You might have formal shoes, but forget to pack your sandals, or you might bring only sleeveless tops and find yourself freezing in air-conditioned buildings. You

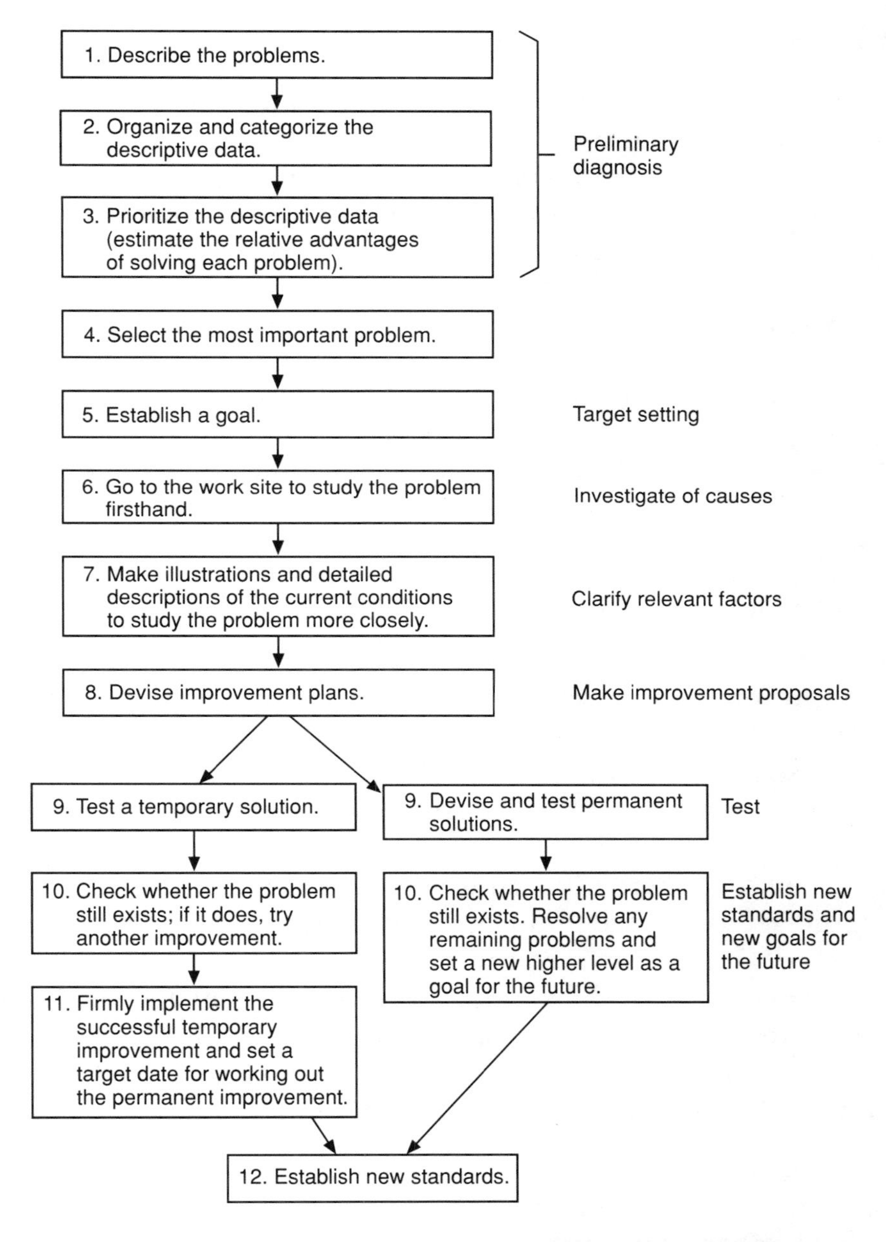

Figure 2-5. Standardization Sequence for Solving Problems

need a more thorough list, including clothing for a range of circumstances as well as specific toiletries, such as sunscreen or aspirin.

2. *Investigate the facts to get to the real problem.*

Use the 5W1H method. Go to the work site and look closely at the current conditions and all of the relevant facts. Keep asking "Why?" until you get to the root cause of the problem. For this type of situation, the Toyota production system prescribes asking "Why?" at least fives times, then, after you have found the root cause, ask "How" to eliminate the cause.

Norio Yanagida's book, *Fear at Mach 1*, describes an investigation into the cause of an airplane crash on Mt. Fuji. The book describes how investigators found evidence that the real cause was unpredictable wind currents near Mt. Fuji. The investigators studied 8-mm film footage of the crash site several dozen times before they discovered the evidence in the film. There had been many different theories about the cause of the crash, but ultimately the film yielded the facts. Through perseverance, the investigators got to the bottom of the matter.

3. *Brainstorm and test specific improvements.*

Many proposals can be made for each goal that has been established. When brainstorming for proposals, it is important to encourage candor and debate. Proposals should be evaluated according to the QCD factors as well as other criteria such as safety and operability. Once a proposal has been selected, it is important to remain flexible and to add to the proposal as needed to overcome problems encountered during the proposal's implementation.

A group once gathered to brainstorm solutions for a certain problem. At first, everyone felt pressured and was reluctant to speak up; the discussion was going nowhere. Finally, one member of the group broke the ice. She held up a Japanese rice ball (a candy filled with sweet bean paste) and squeezed it until the filling started coming out. As she did this, she said, "This is not the best way to get something out in the open!" She then went to the blackboard and wrote out the following "four rules of brainstorming":

- Don't be shy.
- Go for quantity, not quality.
- Build on other people's suggestions.
- Do not criticize ideas — any idea is okay.

This got everyone to relax and speak up, and soon many proposals were on the table, including a very good one that was eventually selected for implementation. Before implementing it, the group looked into the various problems they had predicted, which helped the implementation go smoothly.

GOOD STANDARDS MANUALS HAVE CLEAR CONTROL POINTS

One step not mentioned in Figure 2-5 is the standardization of high-priority items, or determining which problems are most critical and therefore most in need of standardization. Using the example of making a standards manual, we will view prioritization from various angles.

Standards are important as clear statements of who uses what and for what purpose. A standards manual should provide the information that people need, in a form easily understandable to all who use it. A good standards manual should have the following characteristics:

1. *Easily understood descriptions.* The terms, symbols, and diagrams should all conform to in-house rules and conventions.
2. *Consideration of safety, health, and environmental protection.* The manual must satisfy all human and environmental safety standards.
3. *Good interface and compatibility with other departments.* The manual should be written so that it can be understood and used by all related departments.
4. *Appropriate approach for its objectives.* The manual should be oriented clearly toward maintaining high product quality and equipment performance.
5. *Adaptability to change.* The standards should be adaptable enough to require only slight revision when models or operation methods are changed.*

Standardization obviously entails adhering to standards, but this adherence can range from high-priority *must* items to low-priority *preference* items. Standards should describe only those items that are *musts* for getting the job done. You should avoid standardizing methods that are simply a matter of preference or personal discretion and do not really need to be done in a certain way.

---

* Mitsuru Toki, et al., *Easy-to-understand In-house Standardization* (Tokyo: Japanese Standards Association, 1987).

Musts are items that simply have to be checked, because problems in them could lead to defective products, machine breakdowns, or safety hazards. Some examples include temperature and dimension checks during heat processing.

Preferences include items that can be checked or recorded but are already included elsewhere, so they are not essential here. Such unnecessary checking or recording should not be included in the standards manual commonly used at the work site.

Figure 2-6 lists several types of standards. Look at the descriptions of standards in this figure and ask yourself whether their checkpoints should go on a *musts* list or a *preferences* list. Remember, your goal is to make standards that help work proceed more easily and quickly.

As another exercise in prioritizing, imagine that you are including technical and process standards, as described in Figures 1-10 and 2-6, on a one-page operation standards sheet. Which of the following items would be *musts* and which merely *preferences?* Imagine the conditions at your work site and enter *M* for must or *P* for preference after each item (suggested answers below).

1. Manufacturing Division ( )
2. Date ( )
3. Part number ( )
4. Part name ( )
5. Process name or number ( )
6. Materials ( )
7. Machine name ( )
8. Machine number ( )
9. Inspection stamp ( )
10. Operator name ( )
11. Operation standard issue date ( )
12. Work-safety precautions ( )
13. Names of jigs and/or tools ( )
14. Standard dimensions ( )
15. Use time for processing bits and blades ( )
16. Motor sound check ( )
17. Defect count ( )
18. Cutting speed ( )
19. Company issuing the standards manual ( )
20. Standard time ( )
21. Actual processing time ( )
22. Memo for next process ( )
23. Defect causal analysis results ( )

How does it look? Odds are that you entered *M* for a good number of these items. Please reconsider why you believe your *M* items are necessary. After all, aren't some of them already recorded elsewhere and therefore unnecessary? Consider the following points.

Suggested answers: 1-M, 2-M, 3-M, 4-M, 5-M, 6-M, 7-P, 8-P, 9-M, 10-M, 11-M, 12-P, 13-P, 14-M, 15-P, 16-P, 17-M, 18-P, 19-M, 20-M, 21-M, 22-P, 23-P.

Type	Description
1. Regulations	These are formally established task management methods (job regulations, task regulations)
2. Standards	These are product quality requirements based on production standards specified by customers and adopted as in-house standards for products and inspection procedure
3. Specifications	These are restrictions and other conditions placed on suppliers of equipment and parts. Usually, they are discussed and agreed on during supply contract negotiations.
4. Technical standards	These are the detailed standards concerning manufacturing methods and products. They stipulate dimensions, temperature, ingredients, strength characteristics, etc.
5. Process standards	These describe work procedures (processes). They usually appear in work procedure sheets or work instruction booklets.
6. Manuals	These are handbooks used for training and for detailed descriptions of work methods. They also define the company's standards and their objectives.
7. Circular notices	These notices inform people of new or revised standards, necessary preparations or responses, and other related matters.
8. Memos	Memos are a common means of communication for prior notification of extraordinary measures, temporary revisions, or other standard-related matters. They are also used for other types of notices, such as meeting minutes or in-house reports.

Figure 2-6. Types and Purposes of In-House Standards

1. Save time by keeping standards in computer files, to avoid rewriting them by hand when checked or described.
2. Move items that do not need checking at this point to another standards sheet.
3. Organize the items into categories and mark those items that have been checked so far.

Generally, it is not possible to avoid listing items that relate to checking recently processed workpieces; technical standards and quality assurance checkpoints are *musts*. Process standards and ordinary graphs and charts, however, generally do not need to be included.

Since making standards for informing others of abnormalities is difficult, it is usually best to make detailed descriptions of the abnormal conditions on a separate sheet to keep the standards clear and organized. In ordinary lot management, you could apply the above approach when making a standards sheet for individual products that are moved through a series of processes. Figure 2-7 shows an example that combines an item sheet with work instructions and a process progress report. The process control department creates this sheet; the operator fills it in with results of production. The supervisors read it, then send it to the next process to inform them. It is routed in an envelope along with inspection sheets from the QA department and drawings and diagrams from the design department.

STANDARDIZATION FROM THE USER'S PERSPECTIVE

At present, much human work at factories is related to controlling and checking factory operations. Clerical work is increasing, which many factory workers do not welcome. How can standardization help reduce wasteful and unnecessary clerical work?

Standards must be adhered to and records kept to show actual adherence. Such records should cover all of the *must* items and should be designed from the user's perspective as easy-to-use lists for checking against the relevant standards. Record keeping for *preference* items is needed in some situations but not in others.

The concern here is to create user-friendly standards and record keeping so that the standards will be used. The following cases offer examples of this.

Case 1: Improving the Gauge Makes Work Easier

This heat-treatment process has a chamber that should be maintained at 600° (± 6) Celsius. The worker in charge of this process has no temperature gauge and instead uses a millivolt gauge and converts the millivolts to degrees Celsius using a hand-held calculator. Sometimes, a mistake in calculation leads to incorrect temperature settings. This checking procedure is not the worker's favorite part of the job.

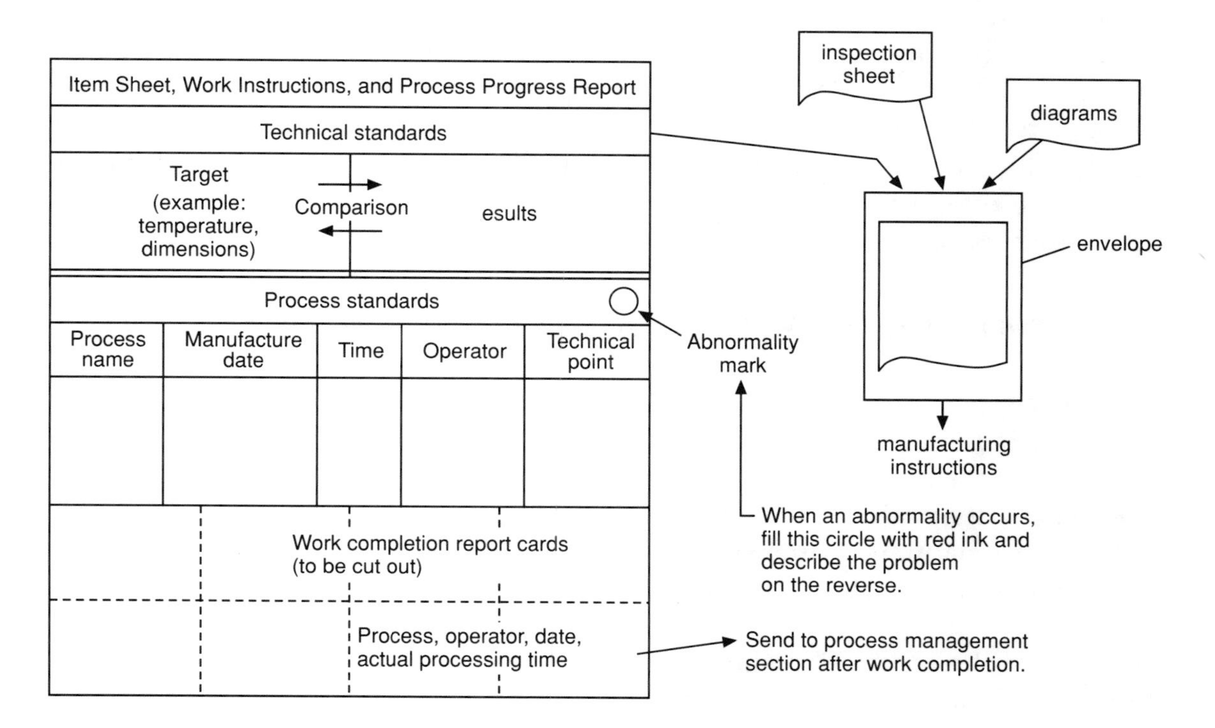

Figure 2-7. General-Purpose Work Instruction/Progress Report

There are two good ways to resolve this problem. One is to keep track of the chamber temperature based solely on millivolt readings; the other is to exchange the millivolt gauge for a temperature gauge (the latter method better serves the goal of user friendliness).

Case 2: Checking Pneumatic and Hydraulic Flow Rates

A worker checks the pneumatic and hydraulic flow rates by noting the flow rate between points A and B and the difference between points A and C. In both instances, the worker checks the values on the gauges, writes them down on the standards sheet, and evaluates the process as defective or nondefective. Misreading gauges and misrecording their data are common human errors in factories, and can lead to waste or even serious safety hazards.

Color coding the ranges on the gauge and also using orange and red to highlight the sections of the form that describe gauge problems improve the checking procedure. Only values in the abnormal range are written down; normal values get a simple check mark. Whenever information appears in an orange or red section, the relevant people are contacted so that countermeasures can be devised and put into effect quickly. This method is much less wasteful and more effective than simply writing down all measurement data from checking procedures.

Case 3: Data Entered on X-R Chart Instead of on Standards Sheet

Previously, workers had to read temperature and current meters, then enter measurements on the standards sheet to confirm normal operating conditions. Since the purpose of recording these measurement data is to discover abnormalities that can occur when jigs and tools are replaced, the data are now recorded to maintain a time-series analysis of the temperature and current conditions. An X-R chart (a quality control chart that expresses individual values and range) is hung on the wall near the meters so the data can be entered directly onto the chart. This eliminates the unnecessary intermediate step of entering data onto a standards sheet.

Case 4: Response to Alarms

The equipment used in a certain process loses hydraulic pressure whenever the oil temperature rises above a certain degree, so the oil temperature must be checked and recorded daily.

As an improvement, a device that detects oil temperatures in the overheated range replaced the data-recording procedure. When the oil overheats, the device lights an alarm lamp and shuts off the hydraulic equipment. This emergency device has proven very effective in preventing equipment operation at low oil pressure, which had been a major cause of product defects.

Since the solution to the alarm condition was usually the replacement of an oil filter, the improvement team also designed the words "oil filter" into the alarm lamp. This reminder made the prescribed countermeasure evident to everyone. Whenever an oil filter replacement did not solve the problem, the workers were advised to get help from an engineer.

Case 5: Treatment of Items Whose Quality Is Difficult to Judge

Before the improvement, inspectors had to distinguish visually between defective and nondefective painted parts. Since paint quality depends on subjective factors that are difficult to judge, some defective items ended up in the nondefective pile while some perfectly good items were rejected.

As a response to this problem, the company decided to establish another post-inspection pile for *possibly* defective items, in addition to the piles for clearly nondefective items and clearly defective items. Items that fell into this intermediate pile got a closer inspection to determine their status.

Although the kind of inspection in this example is difficult to describe in standards manuals, the difficulty was simplified by adding the third pile.

Case 6: Changing the Equipment-Checking Sequence

At this process, no particular sequence had been established for routine equipment checks. To improve this situation, the team figured out the most efficient (least wasteful) walking route for checking the equipment and the most efficient item-checking sequence. To simplify and speed up internal checking, they also replaced screw-on covers with quick-release covers.

The team also made separate lists of specific items to check while the equipment was not running and items to check while the equipment was running or had stopped. They revised some checking procedures to make them more straightforward and logical. For example, abnormal noises from motor bearings were difficult to hear except during start-up, so they devised ways to make such conditions easier to check.

As these examples show, standardization is not limited to what gets written down in standards manuals and forms. Sometimes standardization means introducing a chart, a meter, or an alarm device. The essence of standardization is not to use any particular method, but to make work easier, more reliable, and more efficient (quicker). Doing this well requires you to apply your ingenuity and experience-based know-how.

Advantages of User-Friendly Standards

When you develop standards that are easier to use and adhere to, several benefits result.

1. *Lower costs* due to fewer defects, less overtime, conservation of materials, and elimination of waste.
2. *Prevention of delivery delays* due to fewer sudden equipment failures, fewer defects, and more reliable production operations that proceed as planned.
3. *Lower inspection costs* due to progress in building quality into products at each process, fewer items requiring inspection, and fewer customer complaints.
4. *More efficient operations* due to the creation of a work environment where anyone can easily learn how to turn out nondefective goods efficiently and reliably.
5. *Improved employee morale* due to pride and pleasure in participating in improvement activities, which boosts enthusiasm, self-confidence, and ability.
6. *Improved employee skills* due to the establishment of an easier, straighter path from the novice to the veteran stages of skill acquisition, and the freeing of skilled workers to apply their abilities in more personally fulfilling work.

3
Changing Your Approach to Standardization

This chapter presents several approaches to standardization. Examining these approaches can provide hints for developing a standardization method that is both effective and easy to implement.

GOOD METHODS FOR MAKING AND CHECKING STANDARDS

Dealing with Poorly Prepared Standards

When new standards sheets or work instructions have been written and are about to be used, it is common to first provide some training to the people who will use them. In some cases, the training goes smoothly; the trainees practice the procedures and seem to understand everything. But problems can arise later if the company fails to repeat the original training or does not provide enough training time for the new workers. With the hectic pace of business today, it is common to have the previous worker directly train his or her successor.

Several problems can arise when workers use new standards sheets:

- The worker who is teaching a successor may have misunderstood some of the original training and thus may pass along those misunderstandings, which can lead to problems.
- Some important items might have been left off the standards sheet. This may allow workers to omit certain things from their work, which may not be immediately noticeable but may cause problems in the long run.

- People at the work site may misinterpret the meaning of some items on the standards sheet. They may not realize this until a problem develops later. This happens most often when the standard contains a lot of technical terms and/or complex language.

Each of these situations tends to arise when the standards sheet is poorly prepared. The best solution, therefore, is to improve the quality of the standards sheet.

How to Evaluate a Standards Sheet

The most effective way to evaluate the quality of a standards sheet is to use a checklist based on the following principles:

1. *Does it contain all necessary items?* Do not expect people to act on information that has been omitted.
2. *Even if all of the necessary items are there, can they all be acted on right away?* Long, rambling descriptions are unlikely to receive much attention or response. Even with the whole story on paper, if it is unreadable, it will be misunderstood and ignored. You must write standards sheets in the style of the reader's perspective to inspire action.
3. *Does the standards sheet clearly describe how to respond to abnormalities?* When a problem arises, the sheet should spell out the steps to take to halt production until the problem is resolved.

Steps for Making an Easy-to-Use Standards Sheet

1. Develop a new standards sheet. Have the workers use it without giving them any prior explanations.
2. Have the workers using the new sheet write down their comments or questions about the sheet's instructions.
3. Carry out test runs for a while. Then repeat the first and second steps, this time using workers who have never used the new standards sheet.

Often employees who use a sheet for the first time, use it differently from the way the writers had planned, so the writers add new descriptive paragraphs with the intent of providing extra training. These extra paragraphs are sometimes more than descriptions; they can turn into time-consuming appeals to get workers to change to the new methods. Rather than resorting to added descriptions and explanations, it is better to write a clearer and more complete sheet initially.

One point of difficulty is that people have different approaches and aptitudes toward learning new work methods. The best standards sheet would be tailored to a particular individual's needs, which of course is not feasible. The best practical alternative to ensure the standards sheet's universal accessibility is to involve the users in the writing and rewriting process, which requires user participation.

I have participated many times in standards-writing groups that have followed the steps listed previously. Although this approach takes some time and trouble in the early stages, the result is a standards sheet users identify with, which leads to a much smoother implementation of the new standards.

Supplementary Steps for Double-Checking

Double-checking unduly increases the work load and should be avoided whenever possible; however, there are cases when the tools used for the first check were not up to standard. Possibly the current tools have been used so long that no one knows for which standards they were originally intended.

A certain amount of double-checking can be useful for uncovering such practices. This double-checking is similar to the cross-checking methods bookkeepers use to check totals. For example, to discover data entry errors in figuring the amount of energy consumed to produce the total production output, we can calculate the energy consumed per hour of operation time.

Today, factory supervision relies more than ever on meters, gauges, and other measuring devices. This reliance has increased the risk of defective production due to blind faith in these devices. Consider the following examples:

- The hydraulic pressure gauge gets caught against a drain and cannot move. The factory workers are unaware of problems with the cylinder pressure or equipment operating speed until they discover the defective products that result.
- Fragments from cracked radiator pipes adhere to the temperature gauge's probe, shielding the probe from some of the heat and resulting in inaccurate temperature readings. This problem could be discovered by checking the temperature increase cycle or the energy consumption data.

These examples illustrate the importance of having a supplementary double check to uncover possible problems that might allow conditions to fall below standard.

Process Operations + Inspection = In-Process Inspection

A familiar concept in TQM is building quality into the product. This approach emphasizes quality control in the process to build more quality into the equipment and products. It has helped many companies sweep aside old notions, such as the idea that "I make the products and they (the inspectors) check them." One manifestation of this approach is the poka-yoke (mistake-proofing) system. There are other ways to incorporate an effective inspection into processes, as shown in the following example.

The cycle time for a certain product's machining process is 2.5 minutes. Every two hours a dimension check is made, using a three-dimensional measuring device. All eight machining lines in the factory receive similar dimension checks.

There are more than ten dimension checkpoints on each product. The three-dimensional measuring device plays an important role because of the product's dimensional complexity, and sometimes dimensional defects are discovered. Investigations into the causes have turned up the following: (a) a program change, (b) a tool change, (c) a machine problem, broken bit, or chip-related obstruction.

In such cases, quickly turning out replacements for the defective products might be effective, but management estimated that this response would create more production and labor costs than profit.

They chose instead to implement the in-process (source) inspection method. This method calls for dimension-checking devices such as block gauges and calipers at each process. The workers use these to check workpiece dimensions after each of the three previously mentioned causes. This method worked very well, and the line stopped producing products with dimensional defects.

This successful outcome would have been impossible without the cooperation of line workers who determined and standardized the dimension checking tools used, and decided where they should be kept and how the inspection cycle would operate. This case provides an excellent example of building quality into products at each stage of the process.

Poka-yoke, random spot inspections, and work-in-process sampling inspections are several methods for discovering quality problems within production processes. Whatever method is used, the important thing is to determine whether changes in work-site conditions have any effect on the products and to clarify the cause-and-effect relationships when they do. Examples of such changes were cited above — program changes, broken drill bits due to long-term wear, accumulation of cutting chips, and so on. Whenever you can clarify the causes and effects involved in such changes and then standardize the operations to prevent such effects from affecting product quality, you can help reduce defects.

SKILLFUL RULE-MAKING FOR RESPONSES TO ABNORMALITIES

Most companies describe and manage standardization in two categories: ordinary standardization and standardization of responses to abnormalities.

Ordinary standardization is significant to line workers as the means to confirm and prove they have performed their work up to standard. Standardization is also useful as a kind of mirror on the process operations, as an educational tool, and as a means of explaining what goes on in the process.

Standardization of responses to abnormalities is critical; otherwise only veteran workers know how to respond. The lack of standard response measures makes discovering causes of problems much more time-consuming. Often the line workers must seek help from equipment specialists or equipment-manufacturer representatives. The final cost in downtime and employee hours can be formidable.

While being totally prepared for all kinds of abnormalities is impossible, the standards sheet should incorporate the following matters to improve the speed and accuracy of responses to abnormalities.

- Abnormality trends and predictions
- Appropriate responses to past abnormalities
- Responses that require the help of in-house specialists or manufacturer representatives

Skillful Response to Abnormalities

I have seen good results from companies that have adopted the abnormality response methods copy machine manufacturers use in responding to customer service calls. These are very good methods, as they have been carefully developed by companies that compete intensively to provide the best after-sales service to customers.

Response Methods for Copy Machine Abnormalities

1. An alarm sounds when paper jams or when the machine is out of toner. This method enables quick identification of each abnormality (standardized know-how).
2. When the copy machine operator opens the panel for the section where the problem has occurred, he or she finds a sticker listing step-by-step response instructions for that section. Step-by-step instructions are standardized for each section of the copy machine (this makes it easier to trace the source of the problem).
3. When the above measures fail to solve the problem, the operator must contact a specialist. Three options are:

   - Call a departmental employee who has special maintenance training for the copy machine.
   - Call an employee in a different department who has special maintenance training for the copy machine.
   - If all else fails, call the copy machine manufacturer or service agent.

   In addition, the operator should consult the user's manual for the copier. Standardize the steps taken when initial troubleshooting measures fail. Keep the user's manual close at hand.

I have been involved in implementing standardized response measures for factories in the United States at stages ranging from the completion of factory construction to initial production start-up to full-fledged production. In my experience, the step-by-step method described for copy machines works well in U.S. industries, particularly those with high factory-employee turnover rates and limited new-hire training budgets. Making standards easy to understand and follow is essential for ensuring high quality and correct operation of production equipment.

The copy machine example of applying troubleshooting standards is important not only for quality control and production management (information management) but also as part of a system for responding to abnormalities in general.

Learn Troubleshooting Know-How from the Manufacturer

Case Study: Troubleshooting for Forging-Machine Controllers

Forging-machine controllers use servomechanisms to control the supply of hot water to forging machines. When the controller malfunctions, insufficient hot water is supplied to the forging machine. This causes the forging machine's hardening rate to change, creating various product defects.

Although the controllers come with operation manuals, the manuals fail to provide a clear and simple explanation of the maintenance methods or the troubleshooting steps. When problems arise, it is possible to have the manufacturer send a specialist to identify and fix the problem, but this is expensive and time-consuming. The end result is often a reduction of production output.

Although the controller acts up only occasionally, the company realized it still might pay to establish better troubleshooting measures. The factory managers therefore decided to follow the subsequent troubleshooting steps:

1. They hired a specialist from the manufacturer to write a troubleshooting manual.
2. They taught the operators how to use the manual and also added any needed explanations the manual overlooked. They checked to make sure that the explanations were clear and could be understood easily. Following the "Steps for Making an Easy-to-Use Standards Sheet" (described on page 38), they discussed various questions and added more notes to the manual. (They planned to revise the manual later to include these additions.)
3. The factory managers used the manual to practice troubleshooting on actual forging-machine controllers (three units).

   Unit No. 1: Demonstration by teacher from manufacturer. If points arise that are not covered in the manual, they are added.

   Unit No. 2: Students practice under the teacher's direction. Any questions that arise about the manual's instructions are documented and answered.

   Unit No. 3: Teacher observes practice by students. Precautions and other notes can still be added to the manual.

4. The forging-machine group leader practices using only the manual, with no further explanations. Any questions that arise about the manual's instructions are written down. Someone who has not received the training described in steps 1 through 3 does the practice and the review.
5. The managers have the manufacturer's teacher make up a list of practice problems (theoretical troubleshooting situations). The students write a troubleshooting sequence for solving these problems, which the teacher evaluates and corrects. These theoretical practice problems are then added to the applied problems in the manual. The students then work on solving these problems, just as with the previous problems. The manual also contains instructions for parts replacement cycles and ordering parts from suppliers.
6. The manual and the above revisions are edited into their final form and printed. Standards are established for storing copies of the manuals, preparing the tools, and so on. The factory managers used the step-by-step troubleshooting method described earlier for copy machine problems. Because troubleshooting in this case required checking a wide array of devices, the managers decided it was necessary to standardize all of the troubleshooting methods as one set of standards and built customized troubleshooting carts to hold all the tools they might need. These carts are used not only for troubleshooting but also for periodic maintenance. Figure 3-1 describes one of these carts. They also added the following:

   - A label (near the meter) listing the alarm standards.
   - Labels for each part showing its replacement period and its service-life expiration date.
   - Stickers showing flowcharts of troubleshooting steps for problems that can be solved by simple adjustments; these stickers were affixed to main unit panels in appropriate locations.

With the support of these carts, the operators can respond as effectively as a specialist.

Standardizing the rules for responding to abnormalities makes troubleshooting easier, quicker, and more reliable. This improved troubleshooting helps raise the factory's level of technology.

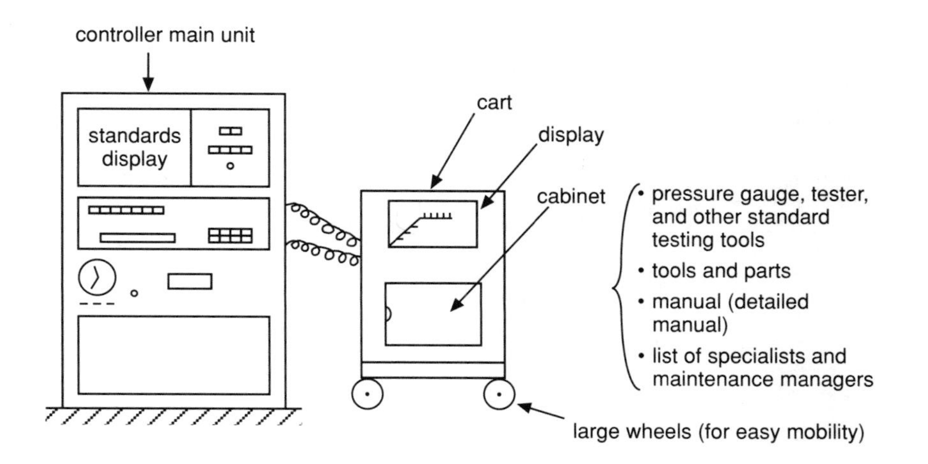

Figure 3-1. Repair and Maintenance Cart for Forging-Machine Controllers

CHARTING A STRAIGHT COURSE TOWARD HIGHER STANDARDS

Figure 1-13 touches on the idea of raising the level of standards. Since standards are one indication of a factory's technological level, they should be improved just as technology should be improved. What is the best way to raise the level of standards? This is the question we will deal with here.

The goal of standardization, which is to enable the factory to make reliable products more easily and more quickly, should be considered first when undertaking standardization. Standards sheets should clearly spell out the method for meeting this goal. These methods should be upheld strictly during the factory's activities. This means that standardization must be preceded by a reevaluation and improvement of how factory employees go about their work. This is a good opportunity to take advantage of the problem-solving steps used in improvement activities developed within the framework of three main improvement methodologies: industrial engineering (IE), quality control (QC), and value engineering (VE). These methodologies have benefited from refinements made over a long time, and as a result they are logical and easy to use. Figure 3-2 shows an outline of their problem-solving steps.

Standardization comes in after the object of improvement is analyzed using the methods described in Figure 3-2 and after the problem is solved; it

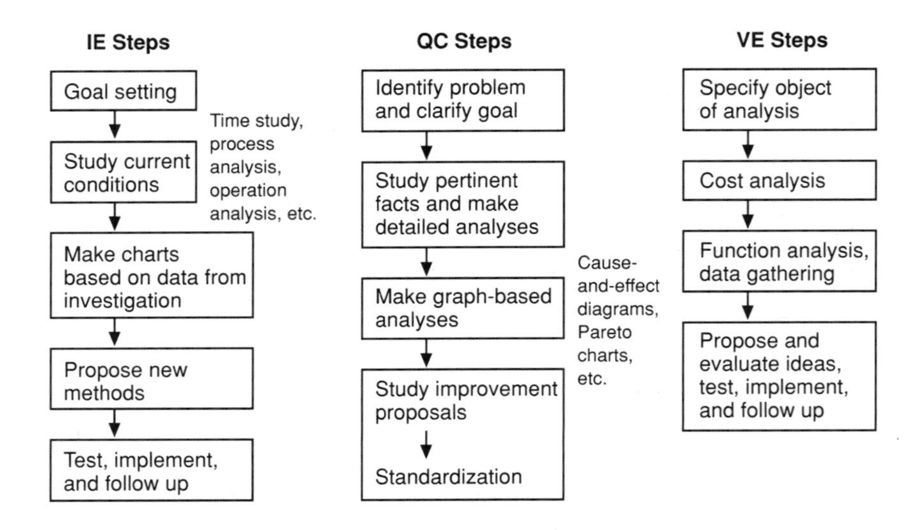

Figure 3-2. Problem-Solving Steps in Three Main Improvement Methodologies

helps make the solution more effective and durable. The following examples demonstrate the essential points in the use of these methods.

Combining Three Main Methodologies to Standardize Zero-Defect Improvements

At a steel-plate rolling process, the operations include rolling a thin layer of paper between the steel plates to help protect them from damage and rust. The same paper is used repeatedly during the several rounds of rolling operations. When the paper tears during a rolling operation, the rolling machine is stopped so that the paper can be repaired (the cost for new paper each time or for a stronger material would be prohibitive), then the rolling operation is restarted. Such paper tearing is therefore a problem not only in terms of equipment downtime, but also because it creates a bottleneck in the production flow that precludes the introduction of automatic or accelerated rolling operations. I was called in to help solve this problem.

I went to an area where the paper repairs were done and saw that the workers were using glue and driers to repair the torn paper. Standards had been made and were observed with regard to the methods for applying the glue, the type of work table to use, and other details. In addition, weekly follow-up meetings were held to discuss such matters as the frequency of paper repair downtime.

But, since our aim was to achieve a higher level of standards — in this case to reduce paper tearing to zero — the current situation of several paper tears per day was clearly intolerable.

They had already tried applying stricter control of the tensile force on the paper spool than on the steel rollers and had looked into improving the quality of the paper material. We decided to conduct QC-style analyses of the pertinent facts. We took samples of torn paper and studied them to determine the causes of tearing. Eventually, we ascertained that the tearing usually started from a point that had been previously glued.

Next, we applied IE methods to make detailed analyses of the tools used in the paper repair operations. We found a variation both in the a-mount of glue applied and in the amount of heat put out by the dryers. As a result the glue did not always dry completely. We were able to confirm this through testing.

These findings prompted us to investigate other methods for applying glue to the torn paper. Here, we used VE methods to study the work of drying glue on sections of paper (formally termed "application of heat to adhesion points"). Since dryers are not the only way that heat can be applied, we investigated and experimented with various drying methods. We found that iron drying was the best method, since it spreads the glue evenly and uniformly applies the same temperature and amount of heat to the glue.

In this example, all three methodologies — QC, IE, and VE — were used to move closer to the goal of zero paper tears.

The VE methodology is not the only one that includes brainstorming for improvement ideas. The same solution could have been found using the ECRS method, a brainstorming approach illustrated in Figure 3-3. Familiarity with all three methodologies is most important in selecting the most useful approach for each situation.

As indicated in its acronym, the ECRS (eliminate, combine, rearrange, simplify) approach begins by asking whether the operation that includes the cause of the problem can simply be *eliminated*. If not, the improvement group should then ask whether the operation can be improved by *combining* or *rearranging* its elements in a new way. If this is still insufficient, they should try to *simplify* (or mechanize) the operation.

The glue-drying example uses the "R" option. The idea of using an iron as the drying method would probably have occurred readily using the ECRS approach. The following example shows another application of the approach.

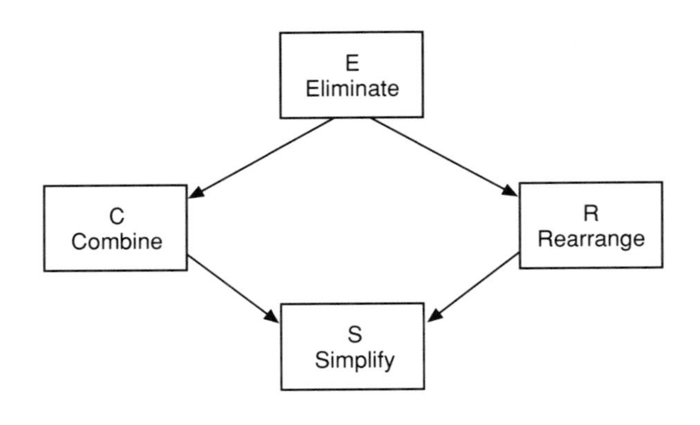

Figure 3-3. ECRS Approach for Brainstorming Improvement Ideas

Using IE Methods to Eliminate the Source of the Problem

This operation improvement example concerns a workshop where an improvement was sought in the way certain parts were processed. The improvement team looked at the work done by an employee who delivered various processing materials to the production equipment and found that he was kept quite busy supplying materials to 30 machines. Without him, the processing equipment would come to a halt. It did not take long to spot the problem in the current operations. This operator was so busy and had been given so much responsibility that he barely stopped for lunch and tended to work overtime in the evenings.

Clearly the work of conveying materials could not be eliminated altogether. Using industrial engineering improvement methods, however, the team was able to reduce the worker's materials conveyance work to the point where he could finish a day's worth of work in about 30 minutes.

Previous work method: The parts supply operator delivered small lots of parts to each machine. To prevent the supplies from running out, he had to constantly check how much was left at each machine and keep supplying the machines, with few breaks.

Improved work method: First, the team determined how many parts each machine needed per hour. They began supplying bigger lots to reduce the number of conveyance trips ("combining"). When necessary, they used a ceiling crane to move the larger containers of parts.

Eventually, the team managed to eliminate even the 30 minutes of conveyance work required of the parts-supply operator. They did this by having the ten line workers get their own parts supplies during changeover ("rearranging"). Thus, what had been 30 minutes of work for one person became just 3 minutes each for ten people, which was not long enough to seriously interfere with their other work. New standards were written to enforce the new operation methods, which had been much changed from the previous methods.

The previously overworked conveyance operator was transferred to a different job, where he is now applying his energy and skills more effectively. These are just two improvement examples, but there is still more to standardization: Before standardizing, it is important to reevaluate the factory work operations and the way goods and equipment are used, then solve whatever problems you discover in the current system. The following section addresses this aspect of standardization.

Determining the Current Level and the Direction Leading to a Higher Level

If you simply implement standardization without first making improvements, you will only establish more firmly the problems that exist in the current system. The correct approach to standardization begins with setting a course for higher levels of operations and standardizing improved methods rather than current methods. Current standards inevitably reflect a limited stage of scientific and technical progress and restrictions on production that existed at the time they were created.

How do you set a course for higher levels? Figure 3-4 lists five evaluation stages and judgment criteria I have developed for this purpose as part of the SN method.* This approach enables us to see how far technical progress has come at each factory and to evaluate its current level.

The chart in Figure 3-4 includes most of the objects of standardization in production lines; it can be broadened to cover other areas as well by adding such categories as information management and energy conservation.

The emphasis here is on the way these methods are integrated. The various categories of improvement shown in the chart are mutually supportive, and standardization should take advantage of the synergy that results when improvements are made in several categories at once. This kind of integrated,

---

* The SN method or approach refers to the author's improvement techniques that incorporate elements of QC, IE, and VE, as appropriate for the particular application. Please see the Preface for a fuller description. — Ed.

Category / Level		Lowest 1	2	3	4	5 Highest
Item	**Appropriate Method**	**Methods that Depend on Veteran Workers**	**Operation Methods Improvement, Related Steps**	**Handling Methods Improvement — Jigs, Tools, and Instruments**	**Production Equipment/Systems Improvement**	**Manufacturing Methods and Materials Improvement**
Quality	Quality control (QC)	• Oral announcement of precautions • Thorough training	• Provision of standards manuals • Visual management	• Poka-yoke devices • Alarm system	• Source control • Self-diagnostics using a computer	• Changes in materials • Changes in production methods • Changes in design plans
Safety	Safety awareness training (SAT)	• Repeated announcement of safety precautions • Confirmation of indicators • Training	• Safety shelving • Safety standards manuals • 5S activities housekeeping	• Auto stoppers • Sensors	• Cutoff system • Guard devices	System changes
Equipment	Preventive/ productive maintenance (PM)	• 5S activities (housekeeping) • PM training	• PM during routine inspection • MTBF measures	• Alarm system • Diagnostic system • Improvement of functions in parts	• Changes in operation methods • Improvements in reliability	System changes
Operation methods	Industrial engineering (IE)	• Training and practice drills • Operations by veteran workers	• Control of operation manual writing process	• Development of operator-machine systems • Low-cost automation	• Computer-based automation and control	Development of FMS, FA, and CIM
Lead Time	Process time (PT)	• Oral communication system • Management without standard times	• Standard-time-based management • Development of Plan-Do-Check system	• Visual management of abnormalities • Auto conveyance at production processes	• Computer control of all processes	Development of FMS, FA, and CIM

Figure 3-4. SN Evaluation of Current Standards Level

broad-based standardization is an important direction for companies to follow in raising their level of standardization. The categories shown in Figure 3-4 are the basic categories needed for moving in this direction.

Figure 3-5 shows the technical level at which the poka-yoke (mistake-proofing) approach serves as a quality control technique. Figure 3-5 also illustrates the role of poka-yoke and other improvement methods in an integrated approach to technical progress.

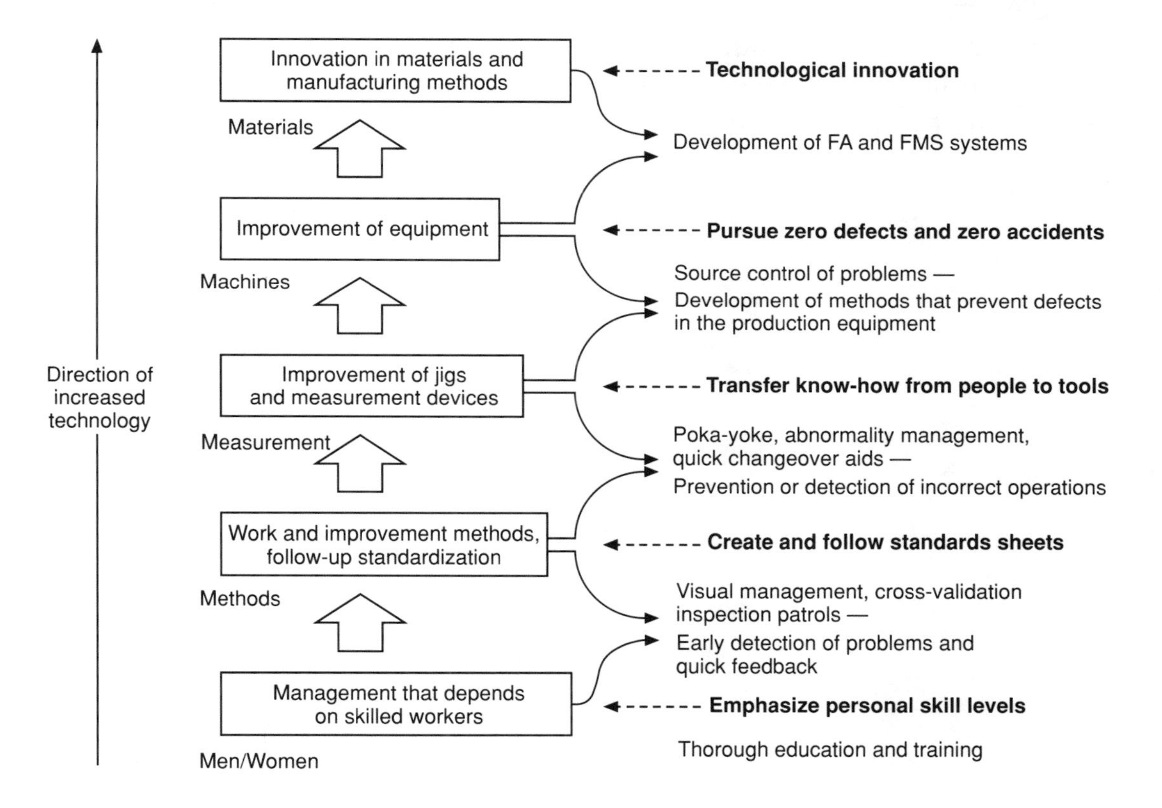

Figure 3-5. Raising the Level of Quality Control

In using poka-yoke, be careful in setting up alarm systems. Some factories have so many lights, bells, and whistles attached to their equipment that the production line sometimes sounds like a brass band. Alarm systems are useful, but they outgrow their usefulness if they are so intrusive that they cause confusion in the factory.

Such has been the case in many factories. In fact, the preponderance of alarm devices has led some factory managers to stop using them and establish some other set of rules for responding to abnormalities. This is a wasteful duplication of effort and resources. To avoid such waste, undertake factory improvements one step at a time, attacking the cause rather than simply building in alarms.

Using the methods and categories shown in Figure 3-4, you can respond to everyday practical problems by selecting what appears to be the most appropriate method for the situation at hand. This is a very helpful reference, since it can prevent factory managers from wasting money (and developing ulcers) in cases where they choose inappropriate responses to problems.

4
Types of Standards and How They Are Made

The previous chapters have covered introductory matters, such as methods for organizing and adopting standardization. This chapter begins to look specifically at the nuts and bolts of standardization.

WRITTEN STANDARDS

Standardization usually starts with written standards. As you reach higher levels of standardization, you tend to use fewer written standards; Figures 3-4 and 3-5 showed some reasons for this. No matter how high the level of standardization, however, you still depend to some degree on written standards. As factory operations become increasingly automated, the written standards become the basic information used by the computers.

The main types of written standards were introduced in Figure 2-6. These include:

1. Regulations
2. Quality standards
3. Specifications
4. Technical standards
5. Process standards
6. Manuals
7. Circular notices
8. Memos

Other common methods for communicating standards, such as vouchers, are also types of standards, but generally fit easily within these eight categories.

This book focuses mainly on the three most frequently used types of standards: the technical and process standards that comprise most standards sheets, and the standards manual. We will look closely at the way these standards are made and used.

Steps in Making Technical and Process Standards

Looking at some examples can give a general picture of what technical and process standards are. Figure 4-1 shows the technical standards used to set the cutting conditions in a machining workshop. Figure 4-2 shows process standards used by operators to learn the correct operation sequence.

Standards sheets should be only one page whenever possible. This makes it easier for the user to read and use the standards.

<table>
<tr><th colspan="10">Coarse Lathe Cutting
Operation method: Lathe cutting   Cut section: Outer perimeter   Material: SUJ2 (annealed)</th></tr>
<tr><th rowspan="2">Electric motor (kW)</th><th rowspan="2">Depth of cut (mm)</th><th rowspan="2">Feed distance (mm/rev)</th><th rowspan="2">Cutting speed (m/min)</th><th rowspan="2">Cutting tool tip (JIS)</th><th colspan="2">Blade angles</th><th rowspan="2">Chip breaker width (mm)</th><th rowspan="2">Nose radius</th><th rowspan="2">Shank size no. (JIS)</th></tr>
<tr><th>γ</th><th>α</th></tr>
<tr><td rowspan="4">2.2 (3HP)</td><td>1</td><td>0.3</td><td>110</td><td rowspan="9">P20</td><td rowspan="9">5</td><td rowspan="9">6~8</td><td rowspan="2">2.0</td><td rowspan="4">0.4</td><td rowspan="4">1</td></tr>
<tr><td>1.5</td><td>0.2</td><td>115</td></tr>
<tr><td>2</td><td rowspan="2">0.1</td><td>130</td><td rowspan="2">1.6</td></tr>
<tr><td>2.5</td><td>120</td></tr>
<tr><td rowspan="4">3.7 (5HP)</td><td rowspan="2">2</td><td>0.4</td><td>90</td><td>3.0</td><td>0.8</td><td rowspan="4">2</td></tr>
<tr><td>0.3</td><td>110</td><td rowspan="3">2.5</td><td rowspan="3">0.4</td></tr>
<tr><td>2.5</td><td>0.2</td><td>115</td></tr>
<tr><td>4</td><td>0.1</td><td>125</td></tr>
<tr><td></td><td></td><td>0.4</td><td>95</td><td></td><td>0.8</td><td>3</td></tr>
</table>

γ: vertical side rake angle α: side clearance angle

Source: Yoshiharu Fukunaga, *Know-how in Production and Process Management Technologies* (Tokyo: Shoryokuka Kenkyusho Corp., 1972).

Figure 4-1. General Model for Technical Standards (Operation Standards for a Carbide Tool)

Process standards sheet

Set on 11/4-10

Item: Automotive part no. A-38		Category	Operation time
		Cycle time:	130 DM
		Net time for Operator A:	81
		Net time for Operator A:	124
Process name: Heat treatment	No. of operators: 2		—
		Extra time for Operator A:	49
		Extra time for Operator B:	6
			—
Notes: • Check hardness of each lot • Weld part onto inner section		Load for Operator A:	37.6%
		Load for Operator B:	4.6%
			—

Operator A	Time
transfer work from conveyor II to oven	23
supply air	32
turn gas switch on	35
turn heating switch on	37
transfer work from conveyor I to conveyor II	41
return to console	60
supply coolant	73
	82
	94

Operator B	Time
remove work from oven	9
transfer work to hydraulic unit	26
transfer work from storage to conveyor I	42
transfer work on conveyor I	60
stop conveyor and sort work	75
inspect inner surface of work	88
transfer work to next process	94
	130

Scale: 20, 40, 60, 80, 100, 20, 40

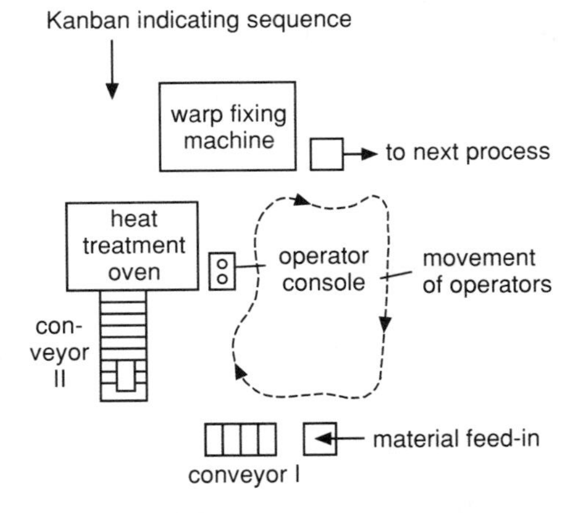

Figure 4-2. General Model for Process Standards (Operation Procedures for Fabrication and Heat Treatment of High-pressure Gas Containers)

The steps for writing technical standards and process standards generally include:

1. Clarify the objectives (what purposes do the standards serve?).
2. List the management items (control points, etc.), which include related data, checkpoints (in sentence form), and other signs and symbols.
3. Divide checkpoints into "musts" and "preferences" categories. Also distinguish between measurement readings within normal operating range (check mark only) and readings showing abnormal operation (record actual measurement).
4. Chart the data along with their control limits. Make data charts easy to use during operations. Detailed data and other matters that cannot be expressed easily in words should be illustrated using photos or drawings.
5. Write the standards sheet in a format appropriate to its purpose.
6. Revise the standards sheet based on input from others, then use the standards on a trial basis.
7. Check the results of the "test run" and, if satisfactory, submit the standards for approval. If there are problems, correct them and return to step 6.
8. Enter the date of issue, the company name, and the name of the manager responsible.
9. Display the standards at the work site. If possible use aids such as color-coded displays or kanban cards to make them more easily understood.
10. Train work-site employees in the new standards and then implement thoroughly.

What to Include in Manuals

One type of manual explains how to use and write standards. The other type, which includes the typical equipment-related manual, explains such matters as motion principles, part structures, part service life, supplier addresses, and troubleshooting procedures. Some examples of items that should be compiled in this type of manual include:

1. *Main title or statement (of the standardization):* The heading should clarify the objective and the purpose of the standard manual as a whole.
2. *Application scope:* This should state the intended range of use and should include the standard's date of issue, department, and author.
3. *Brief preface:* The preface should give references for any related manuals or materials.

4. *Table of contents*
5. *Flowchart:* Some manuals do not include flowcharts; however, along with the table of contents, flowcharts are recommended because they can quickly give people an idea of the requirements described by the manual.
6. *Subtitles:*
   - These should mention the principles, mechanisms, or other matters that are central to each aspect of the work at hand. Alternatively, having a main subtitle that briefly summarizes the contents might be useful for making the body of the text more easily understood. Sometimes, the main subtitle mentions a company policy discussed in the body text.
   - Subtitles should describe specific subelements.
7. *Troubleshooting methods*
8. *Equipment maintenance points, parts replacement periods and service life, and addresses of part suppliers*
9. *Examples:* This section should describe frequently occurring situations where expert assistance must be requested.
10. *Alphabetized index*
11. *Standard document number and approvals:* The standard should be initialed or stamped to show authorization by the factory (or company) management.

Manuals should be written with the same care as textbooks: They should include user-friendly devices, such as a detailed index that enables users to quickly find the subjects they are looking for.

STANDARDIZED OPERATION INSTRUCTIONS

In today's era of rapid diversification and change, factories must produce a constantly changing variety of products, which requires frequent changes in factory operations. Although they see changes in the market and in customer needs, companies have been trying to standardize factory operations and to get mass-production benefits from the same production line design. Effective standardization can lower production costs without raising personnel costs. The group technology (GT) approach has been a very effective standardization method in this regard; however, it involves planning for parts design and processing to produce a maximum variety of parts from standardized operations on a few basic types.

Today, managers are learning to operate wide-variety small-lot production systems and many companies are working toward the goal of one-piece production flow. These changes are radically reforming production lines. A key feature of these changes is that these new types of production lines can be adapted quickly to changes in operation instructions. Below are some examples of these kinds of standardized operation instructions (the examples given are very basic, since operations vary depending on the factory's production system, the production management system's requirements, and other factors).

Specifically, we will examine

1. standardization at the design stage
2. standardization of the production line
3. planning for leveled production
4. production instructions.

This order of standardization follows the order of production flow.

Standardization at the Design Stage

Figure 4-3 outlines an example of the process of standardization at the design stage. To make this kind of standardization efficient, you usually start with a review of the entire product line.

If, however, the product line is too extensive to take on all at once, prioritize the products according to their monetary value until you have accounted for about 90 percent of the product line's total value. Next, group the products according to their similar parts. Once you have grouped the parts in this way, you can begin determining the greatest common measure (GCM) of part dimensions. Each newly designed part shape is a prototype; it is not put into production until an analysis determines whether it can be produced by further processing of an existing standardized part. This parts standardization approach is called group technology or GT.

To make the GT approach easier to understand, consider, for example, how various shrimp dishes are prepared in a seafood restaurant.

The raw shrimp is a GCM part for this restaurant. It is dipped in batter and deep-fried to make tempura, served chilled with cocktail sauce, or stewed with spices and other ingredients to make gumbo. The variable factors here are the tempura batter, the act of chilling and arranging the shrimp cocktail, the stewing with other ingredients, and so on.

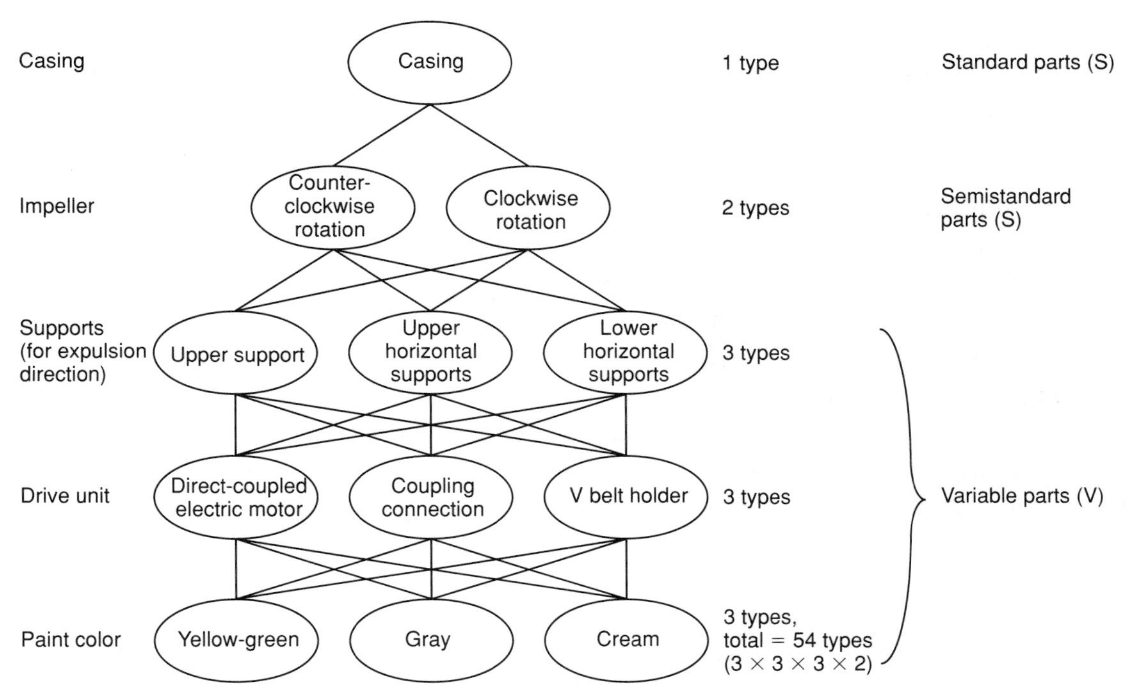

Source: Masao Umeda, et al., *12 Points for In-house Standardization* (Tokyo: Japanese Standards Association, 1984).

Figure 4-3. Product Standardization Using the Group Technology Approach (Air Blower)

Thus, the GT approach categorizes parts into "basic" or "GCM" parts and "variable" parts or processes. Then it systematically standardizes how they are combined when new products are designed. This enables diversification of products while using a smaller variety of parts. The cook's method for arranging the ingredients could be based on a diagram similar to that shown in Figure 4-3. This method seeks to reduce the variety of ingredients (parts) required while serving the customers' needs for diverse product models, whether they are air blowers or menu items.

Production Line Standardization

Besides helping companies design new parts, the GT method helps the production line produce variety efficiently. Production line standardization makes production more efficient in many respects, including

1. Less time required to become accustomed to changes in production operations

2. Less equipment downtime for changeover
3. Less time required for production instructions and production management

Some people might think that GT is just the quick changeover approach under a different name. Although quick changeover is one method that can be used to enable a mixed production flow, it is not the only method. A production line that has already achieved quicker changeover times can still benefit from applying the GT approach. In many cases, the design of production lines has used a combination of quick changeover and the GT approach.

Figure 4-4 illustrates the "lane management" method used by Company H. This method is basically an application of the GT approach, but here, they used their process analysis results to establish four similar processing lines, which they call "lanes," and have established level-load production. Each lane has a

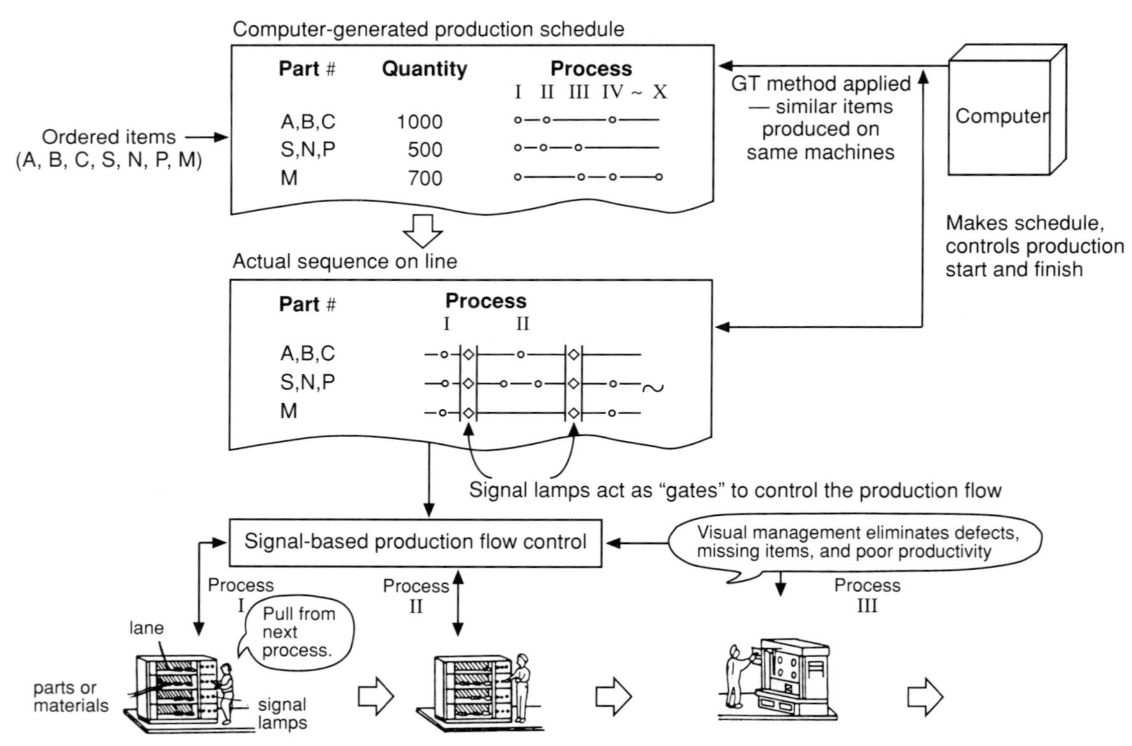

Source: Yuzaburo Sakamoto, "Factory Process Management," *Kojo Kanri (Factory Management)*, 36(2): 73-89 (Jan. 1990).

Figure 4-4. Example of GT Lines at Company H

signal lamp panel (green, yellow, and red) that controls production flow by indicating what to process, in what quantity, and by when.

Planning for Leveled Production

The concept of production leveling (also known as load leveling) became famous along with Toyota's kanban system in the wake of the oil crises of the 1970s. This is one of the basic approaches used to develop just-in-time production with very little waste. Detailed descriptions are available in other works; here, the discussion is restricted to how to plan for smooth production.

Figure 4-5 shows an example of the steps used to develop smooth production based on the current product-order data. When Company H receives product orders from company N, they first categorize the data according to delivery deadline. Next, they analyze the categorized data to determine which items need production and delivery the soonest.

The first step is to subtract from the ordered products any product that the production line already turned out before planning a new line. In this example, the schedule for producing the required output is worked out using a computer. At this stage, you evaluate the required output data for the detailed production schedule to design a production line whose production capacity is well balanced, and to make sure that all the required jigs and production equipment are available. After that, you are ready to work out a detailed schedule indicating which process handles how many of which product by what time.

As shown in Figure 4-5, when the load exceeds the equipment capacity or reaches the peak value, you apply production leveling to adjust the load. Be careful to work out a practical production schedule that enables the smooth handling of the load. The production instructions issued to the work site are based on this practical production schedule. The production line can minimize waste that occurs when adapting to changes by following these instructions.

Although this discussion is general, the key point is that unless production leveling follows these steps and is well understood at the work site, the work site usually ends up operating in a way different from the production instructions.

Along with a schedule created using the steps shown in Figure 4-5, you need production management to ensure that the schedule is carried out smoothly at the work site. Many companies use the kanban system as a method for exercising such production management, and we will briefly examine an example of this approach.

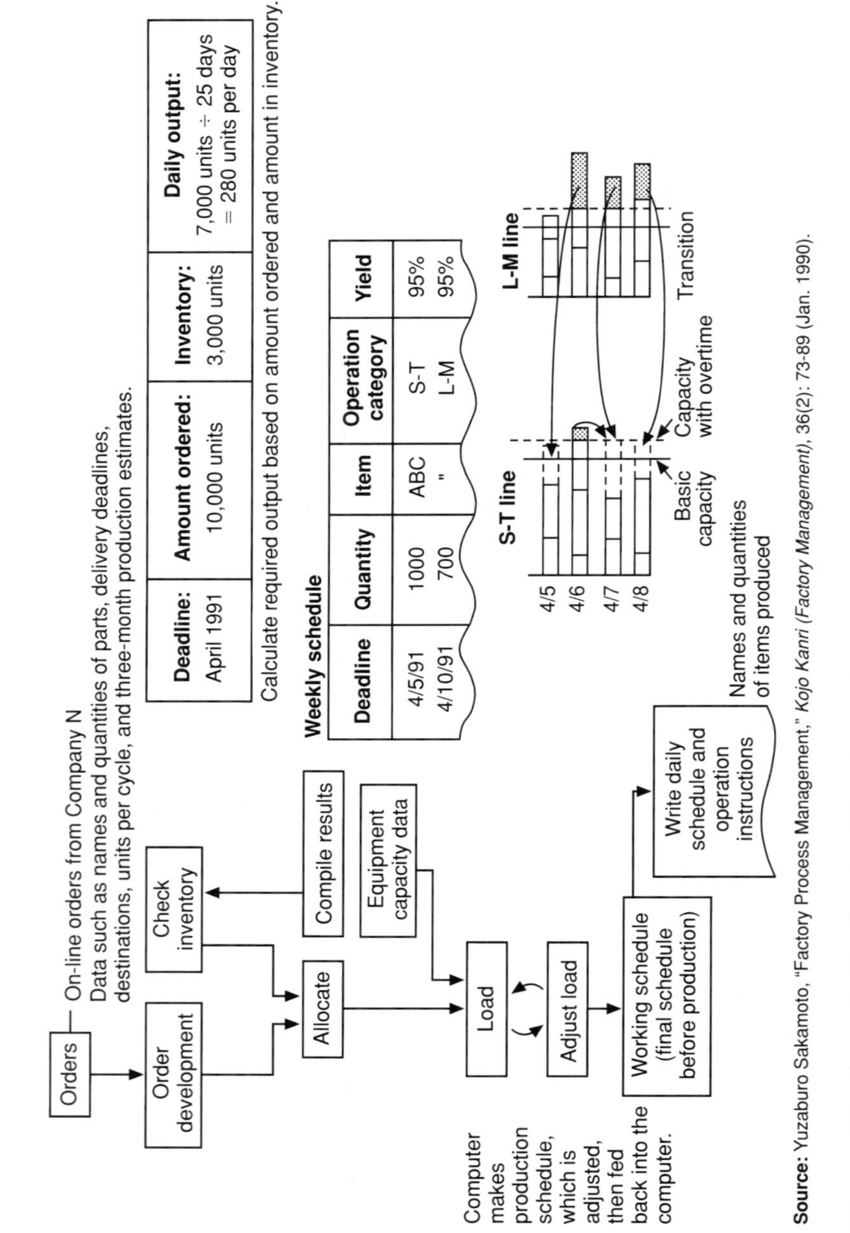

Source: Yuzaburo Sakamoto, "Factory Process Management," *Kojo Kanri (Factory Management)*, 36(2): 73-89 (Jan. 1990).

Figure 4-5. Steps in Production Planning at Company H

Figure 4-6 shows how the kanban are used as automobile production instructions indicating how many units must be produced. These kanban serve as production requests from downstream processes to upstream processes. The parts requested on the kanban are carried along with the kanban to the next process. The method is simple because when the line workers run out of kanban cards for a certain product, they know they have finished with that product.

Three Approaches to Production Instructions

Production instructions are especially important in wide-variety, small-lot production. There are three main ways of planning and implementing these production instructions.

1. Assuming a work site has standards (or work instructions), when requests come in to make parts in certain quantities, the operators check the standards or instructions to see how the parts should be made (many standards sheets have a list of operation instruction points).
2. The preproduction department assembles a set of instructions that includes the names and amounts of the parts needed and all related drawings and work instructions; sometimes even the tools are included. The set of standards is delivered to the work site only when that product needs to be made there.
3. Work instructions are sent to the work site just before the required items need to be made or on a daily basis. The kanban system is an example of this method.

For practical reasons, some companies find it best to use a combination of these methods. Figure 4-7 shows examples of approaches 2 and 3, which are the most common ways. The work instructions are given on a form that includes relevant cautionary points. This method's special feature is that it combines work instructions with specific educational points for the work-site operators.

METHODS FOR EXTRACTING ONLY THE DATA NEEDED, ONLY WHEN NEEDED

The advent of computer-based production management (also known as computer-integrated manufacturing or CIM), has made the technical aspects

Example of kanban label used on final assembly line

Assembly no.			Destination	
Model	AJ56P-KFH			
Rear spring	Rear axle	Booster	Steering lock	Collapsible steering wheel
	S	M	A	
Differential gear ratio	Free wheel hub	Electrical	Emission control	Transmission
400				
Alternator	Air cleaner	Oil cooler	Heater/AC	Front winch
500Z			H	
Antifreeze/oil	High-level compensation	LLC	Fan	Rear hook
			D	
EDC				For cold climate
A				

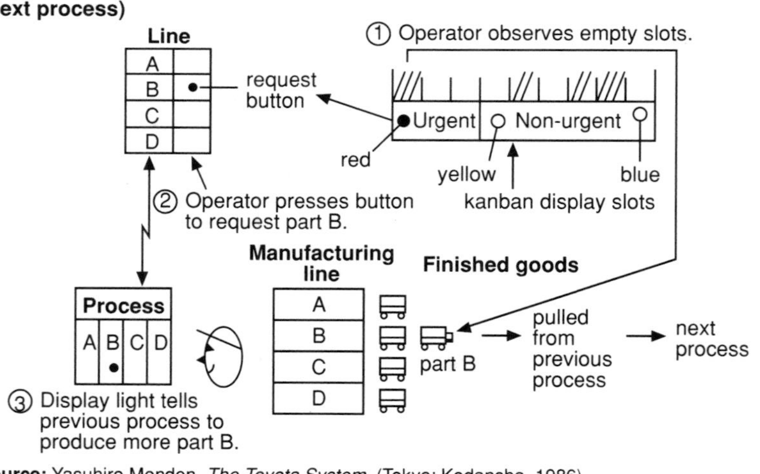

Source: Yasuhiro Monden, *The Toyota System* (Tokyo: Kodansha, 1986).

Figure 4-6. Example of Kanban Use

Work instruction form	**Lathe cutting of XX spindle**	W-3553 1/2
Approval date: Feb. 14, 19—		Lathe work

A. Process

Process	**1**	**2**	**3**	**4**	**5**
Model	TL	LAT	MIL	LAT	GRI
Specified time	0.5 2.5	0.5 0.55	0.4 0.65	0.5 1.1	0.3 0.5

B. Machine to be used
LAT-1744

C. Jigs and tools

Item	**Model No.**	**Units**	**Comments**
A-section gauge	T-12.3456-IU	1	
Length gauge	" -IX	1	
"	" -IW	1	
"	" -IV	1	
Copying die	" -IAP	1	
Temporary center	" -IZ	2 pr. 4 un.	
Screw gauge		1	

D. Processing (Process No. 2)
Remove tag from spindle. After processing, replace tag and check to be sure it is the correct tag.

1. Fit both ends to the temporary center (see Fig. 2), then attach as shown in Fig. 1.

Fig. 1

2. Using the template, process the outer perimeter. When doing this, inspect the finished section and leave 0.4mm extra for polishing. (The R section must be well finished.)

Fig. 2 Temporary center

Inspector: Issue date: Drafted by:	Revision No. ④ ③ ② ①

Source: *In-house Standardization Handbook*, 2d ed. (Tokyo: Japanese Standards Association, 1989).

Figure 4-7. Example of Specific Technical Standards in Work Instructions

of production management easier to handle, lowered system costs, and interested many manufacturing companies in developing a computer-based system that precisely serves their needs. Standardization is the key to custom-tailoring such systems.

The benefits of using computers include the ability to store vast amounts of data from which users can extract data when it is needed. From the work site's perspective, computers are very useful because they give site managers direct access to various kinds of production information. This wide access to data helps prevent misrepresentation of the facts and enables managers to manage more precisely and confidently.

Standardization can be applied to various aspects of computer-based production management systems to create powerful management tools. Although the example shown in Figure 4-8 is titled Computer-Based Daily Production Management, it also shows the special instructions needed at the work site, and the tallying of results. This example focuses on the names and amounts of parts used, the relevant technical standards, and the quality-control results. The forms can also be used to check equipment capacity utilization and other management-related data. Factory automation can be seen as a further technological refinement of a CIM system that eliminates all human labor from the process of distributing work instructions.

ESSENTIAL STANDARDIZATION FOR NEW-EMPLOYEE TRAINING

No one would question the importance of teaching operators how to perform their work correctly; however, this is easier said than done. Training new employees is a major issue for every company. In the broad sense, new-employee training is aimed not only at newly hired employees but also at long-time employees who transfer to new posts that require new skills and training support.

There are two main types of training systems in use today:

1. Employee-to-employee training, which is the most common system and remains a key method for teaching standards to employees.
2. Training by specialists or managers

Visual management tools are also used as training aids, with graphic instructions showing workers what to do in each step of a procedure. Another approach, used widely at Toyota, has the first-line supervisors demonstrate improved work procedures to line operators and then help the operators learn to perform the new procedures smoothly. The objective of all of these training systems is to accelerate the process by which new, inexperienced operators become "veteran" operators. The following sections examine the main points of the three training systems.

Employee-to-Employee Training

Many companies have established some sort of employee-to-employee training system. This is not to say, however, that most of these systems are effective. The standardization theme of making production easier, quicker, and more reliable also applies to standardization of training systems. In fact, the building of a good training system is a basic part of standardization. Accordingly, the standardization approach should be applied in developing the steps used in employee-to-employee training and can be used in other types of training systems.

1. The teaching employee discusses the significance of the operation, its basic points, and all relevant safety concerns.
2. The trainee visits the work site to get a general idea of the work done there.
3. The teaching employee slowly demonstrates the steps of the operation, explaining the key points for making the work easier, quicker, and more reliable.
4. The trainee watches veteran workers perform the operation, then goes back to point 3 for a review lesson.
5. Under the supervision of the teaching employee, the trainee tries to do the work, or repeats the steps performed by the teacher. It is best to start with the easier parts of the operation before working up to more difficult parts.
6. When he or she is confident enough, the trainee tries to perform the operation unassisted. The teaching employee comes by to check occasionally, and immediately gives more training to remedy any mistakes that occur.
7. The teaching employee encourages the trainee to raise any questions or doubts about the work. When both feel the trainee has completely mastered the operation, he or she is left to work unsupervised. After that point, checking takes the form of the supervisor's review of output figures, defects, and operation-rate results.
8. A gathering is arranged to officially welcome the new employee and help him or her feel part of the group. This welcome also helps make the

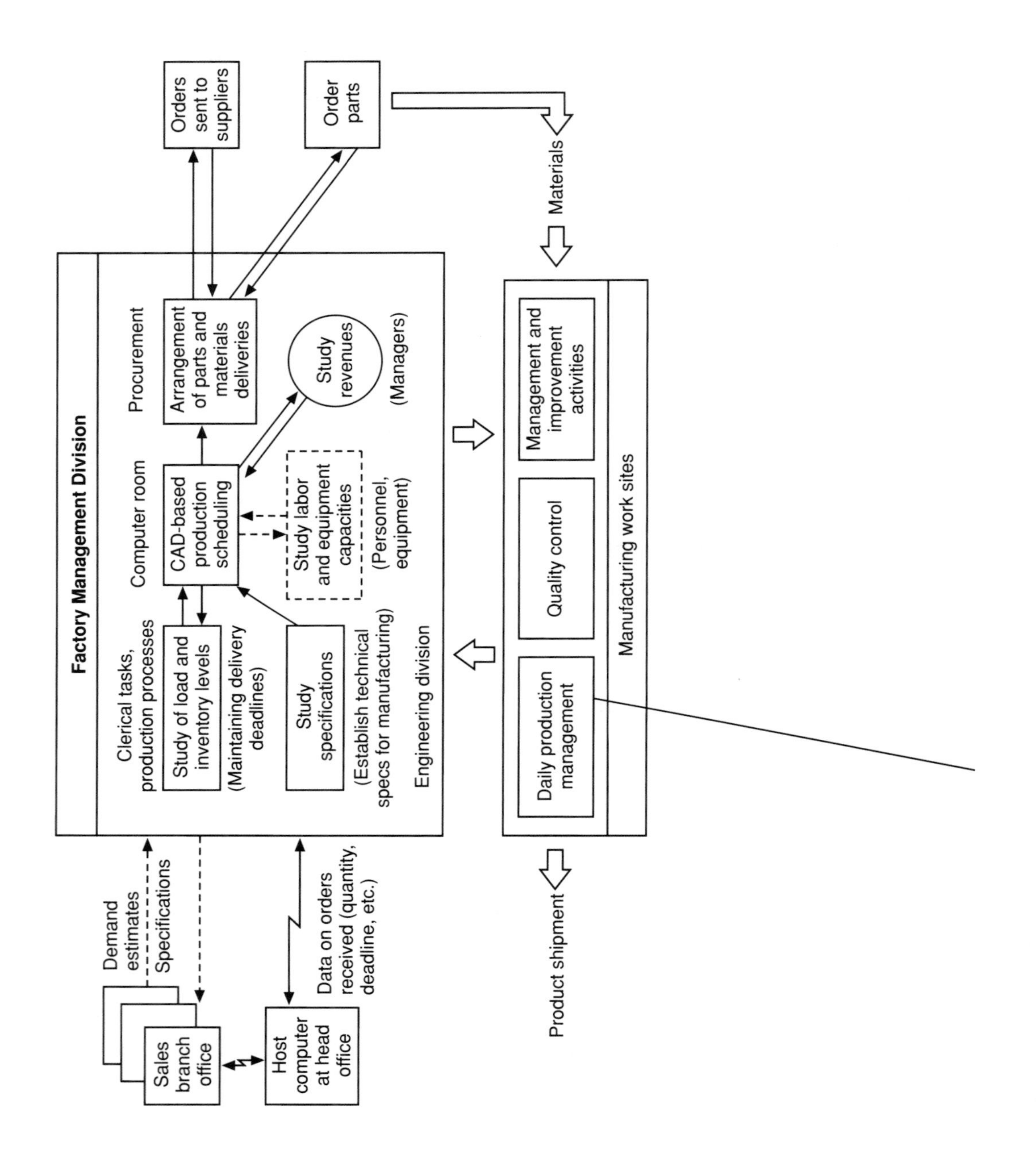
Factory Management Division
Clerical tasks, production processes
Computer room
Procurement
Study of load and inventory levels
(Maintaining delivery deadlines)
CAD-based production scheduling
Arrangement of parts and materials deliveries
Study revenues
(Managers)
Study specifications
(Establish technical specs for manufacturing)
Study labor and equipment capacities
(Personnel, equipment)
Engineering division
Orders sent to suppliers
Order parts
Materials
Daily production management
Quality control
Management and improvement activities
Manufacturing work sites
Product shipment
Sales branch office
Host computer at head office
Demand estimates
Specifications
Data on orders received (quantity, deadline, etc.)

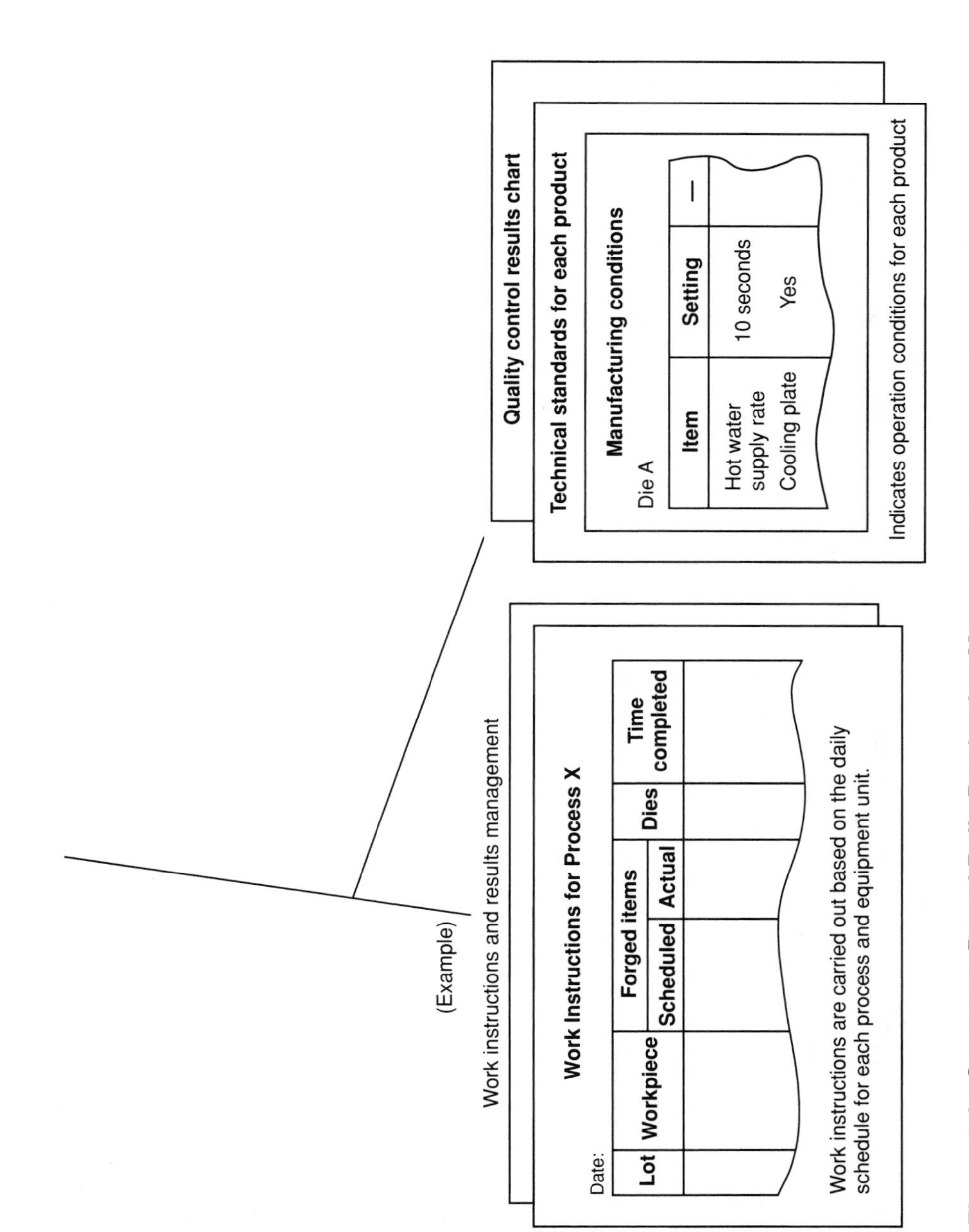

Figure 4-8. Computer-Based Daily Production Management

new employee less hesitant to raise questions or doubts in the future. Friendly, trusting relationships among coworkers are an important element in smooth operations.

9. The new employee is asked to participate in improvement projects and brainstorming sessions. The improvement group may assign a particular problem-solving task to the new employee.

Training by Specialists or Managers

The basic steps of training by specialists or managers are similar to those in employee-to-employee training.

1. The teacher emphasizes the purpose and significance of the operations being taught and gives the trainees a training schedule.
2. The teacher describes personal difficulties with the work in the past and expresses the hope that the operations will go even more smoothly in the future. If the trainees can take a positive attitude toward the work, the teacher's job is half done.
3. The training is hands-on and work-site centered. Trainees are encouraged to ask questions at any time. If there are no questions, the trainees are given a few case studies of problems and asked to raise questions and suggest solutions as a learning method.
4. The teacher makes a point of praising trainees for good suggestions.
5. The teacher quizzes trainees to test their mastery of work procedures, adding review lessons to correct any apparent deficiencies.
6. After confirming that the trainees have mastered the work procedures, the teacher uses a step-by-step approach to teach relevant troubleshooting methods.

Employee-to-employee training and training by specialists or managers both resemble the kind of training that athletes receive from their coaches. Typically, a coach makes out the athlete's training schedule, provides basic instruction, teaches practical skills, and gives extra practice and training support before each game or tournament.

As shown in the multiple skills achievement chart in Figure 4-9, teachers can help trainees become aware of their current skill levels and build their ambition to reach training goals by displaying their status (such as "mastered," "in training," and "untrained") in each type of skill or operation taught. Good

Skill / Operator	1	2	3	4	5	6	7	8	9	
A	○	●		●	●	○	○		●	
B	○		●		●	○	●		○	
C	●	○	●	●	●	●	●	●	○	
D	○			○		○		●	●	
E	○		●		○					

● : Mastered   ○ : In training   Blank : Untrained

Figure 4-9. Skills Achievement Chart

coaches know how to offer criticism and heap praise on their trainees when appropriate. As long as the training methods are sound, training given in an enthusiastic, coaching spirit leads to quicker mastery of the desired skills.

$$\text{Enthusiasm} \times \text{Opportunity} \times \text{Effort} = \text{Success}$$

The key to fast-paced and efficient skills training is standardization to develop systematic training methods that incorporate all three of these ingredients for success.

Training by Visual Management

If the standardization of the operation is clear enough to be described visually, you can use the visual difference between results and standards to point out where defects and other problems lie. This can be an effective training method.

Many companies have already developed visual management systems for improved communication and early detection of problems on the production line. At one company, for instance, measurement instruments such as calipers and micrometers are marked with a different color after each month's routine calibrations (e.g., red for January, yellow for February, and so forth) so that it is easy to see when each instrument was last inspected. Other visual management methods include making charts to illustrate differences between production target values and actual result values, using colors to show where results exceeded targets (green) or fell short (red). This makes it easy to spot current production problems. Some of these visual management methods have long been standard procedure in factories.

In addition to the objectives of preventing and detecting problems early on, visual management also demonstrates the level of standardization attained by the factory or company; using an approach like color coding indicates a high level of standardization. In this regard, visual management is a very useful training tool. Figure 4-10 lists several examples of how factories apply visual management. Not all of these examples fall into the realm of standardization, but they are worth mentioning as practices many companies use successfully.

STANDARDS LIGHTEN THE MANAGER'S BURDEN

Although the style may vary from one management book to another, the message is basically the same: "Management means applying your total resources and abilities to achieve a goal related to creating value." Three points deserve emphasis concerning the proper role of managers:

1. Managers can delegate work to subordinates, but they never delegate complete responsibility.
2. Managers try to build a system that enables their employees to make the most of their abilities.
3. Managers systematically promote the Plan-Do-Check cycle and work toward higher efficiency.

Part of a manager's burden is that he or she is never free from responsibility, even when another employee actually carries out the work. That is why many managers find it difficult to confidently entrust employees to get things done.

Standardization is an essential part of the activity of management, but it can also help lighten the manager's burden. When standards are in place and followed, work goes more easily, more reliably, and more rapidly. Consequently, fewer problems and customer complaints arise regarding quality, cost, and delivery, which lessens managers' responsibilities. The more closely standards are followed, the less responsibility and work managers have.

In this sense, standards can be viewed as the medium by which managers can pass authority to their subordinates. Carefully followed standards relieve managers of the tedious task of checking on operators, equipment, and materials. They make it easier for managers to trust the factory's operations, and thus enable them to devote more energy to other aspects of a manager's work, such as those mentioned in points 2 and 3 above.

Management Target	Implementation Items
1. Process management & deadline management	Production management board (for progress control), fluid curve chart (a graph comparing target values to result values), display boards (e.g., for work instructions, parts delivery schedules, and urgent items or delayed deliveries).
2. Quality control	Defect graph (rate and trends), defective goods storage area, display of defect-prevention rules, defect samples, etc.
3. Operation management	Operating status display lamps: Green = normal conditions yellow = changeover in progress red = breakdown or abnormality Operation standards (technical standards and process standards), multiskills training achievement chart, cutting tool replacement schedule and results chart, and equipment capacity utilization graph; bulletin boards for notices regarding causes of minor line stoppages, improvement campaign results, etc.
4. Materials and parts management	Storage site specifications, part names, kanban card displays, displays of minimum and maximum allowable inventory, floor area and height restrictions for storage areas, and defect storage site indicators; notices about missing inventory items, retained items, items awaiting disposal, items to be repaired, etc.
5. Management of equipment, jigs, and tools	Maintenance schedules and results charts, displays of equipment checkpoints (sections and check items), routine inspection check sheets, storage site instructions for dies, jigs, and tools (including inventory and ordering information), spare parts inventory, inventory shelf management, tags indicating name of equipment manager, notices describing reasons for breakdowns, etc.
6. Safety	Safety guidelines and special safety-related notices
7. Improvement goals and improvement management	Weekly and daily charts showing progress toward improvement targets, equipment capacity utilization results, defect elimination results, work-in-process and warehouse inventory trend charts, safety trend charts, 5S and PM activity progress charts, improvement proposal campaign results, displays of improvement case studies, etc.

Figure 4-10. Examples of Visual Management Targets and Implementation Items

Therefore, standards are intimately linked to a company's chain of responsibility and are a main support for quality assurance activities. Imagine, for instance, that two companies are negotiating a supply contract. To investigate each other's quality assurance systems, it is common practice to inspect diagrams of each company's standards system and the departments and people responsible. A company's quality assurance system shows outsiders an otherwise invisible part of the company known as trustworthiness.

Figure 4-11 is an example of such a diagram showing the relationship between a company's chain of responsibility and its standards. Not shown in the figure are other important quality indicators, such as operation-related (production capacity) documents and QC process diagrams. These are also areas that receive support from corresponding types of standards.

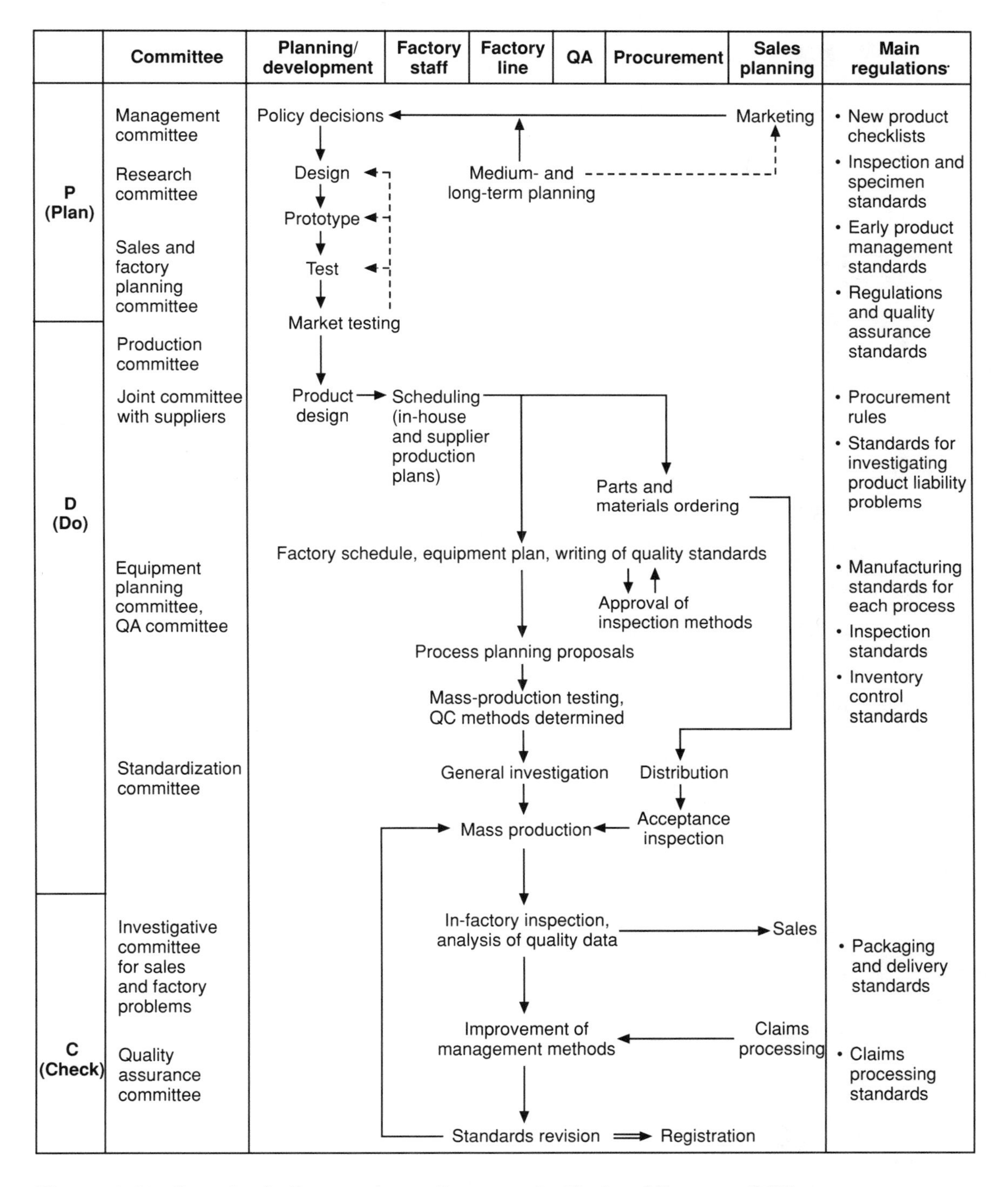

Figure 4-11. Standards Supporting a Company's Chain of Responsibility

5
Standardization for Quality Control

This and subsequent chapters examine the process of building up standardization and explain the process in terms of standardization themes. Reviews of several case studies give a better understanding of the meaning of "raising the level of standards."

As mentioned before, the overall objective of standardization is to integrate the three key factors of QCD (quality, cost, and delivery). Our first theme is standardization for quality control.

WHAT IS QUALITY?

Generally, we think of "good quality" as the opposite of "defective quality." However, quality can be nondefective without being particularly good. What is the basic definition of quality? Armand Feigenbaum, in his book *Total Quality Control*, offers the following:

> In the phrase "quality control," . . . the word "quality" does not have the popular meaning of "best" in any abstract sense. To industry, it means "best for satisfying certain customer conditions," whether the product is tangible . . . or intangible.
>
> Important among these customer conditions are (1) the actual end use and (2) the selling price of the product or service.*

---

* Armand V. Feigenbaum, *Total Quality Control*, 3d ed. (New York: McGraw-Hill, 1983), 9.

Here is another definition:

> Quality is the combination of product characteristics from both the engineering and manufacturing perspectives; when the product is put to use, quality is also the degree to which the product meets the customers' expectations.

These definitions make three salient points:

1. The product should work in a manner that is satisfactory to the customer (i.e., functional quality).
2. The product's cost should be appropriate to its functional quality.
3. The product should be reliable (easy to use, resistant to breakdowns, attractive, easy to maintain, easy to repair, etc.).

The most important quality factor for manufacturers is knowledge of what makes a product marketable, including knowledge of the ways customers use the product. Quality control standards grounded in this knowledge are the source of authority for building quality into products at each production process.

Figure 5-1 describes five factors (the 5Ms) that determine product quality: materials, machines, methods, men and women, and measurements. Processes gradually transform materials into products and build quality into products.

"Building quality into products" means using the 5Ms to eliminate deviation from the relevant standards. Quality control standards are used as a technical means of controlling deviation from quality. The two factors you must know to maintain such standards are the principles behind standards as a technical means of controlling deviation from quality and the characteristics that make a product marketable.

DOES IT TAKE EXPERIENCE TO PRODUCE GOOD QUALITY?

First Factor for Quality Improvement: Understanding Technical Principles

At a cutting machine process, the defect rate was low when a veteran operator handled the equipment, but it soared whenever a less experienced operator was put on the job. By closely observing the veteran operators, the managers found that they occasionally made adjustments to the cutting machine. Less experienced workers also made machine adjustments, but only after making a defective product; their adjustments tend to take more time and were less effective. The general attitude at the factory was, "Well, that's what makes the difference between veterans and novices."

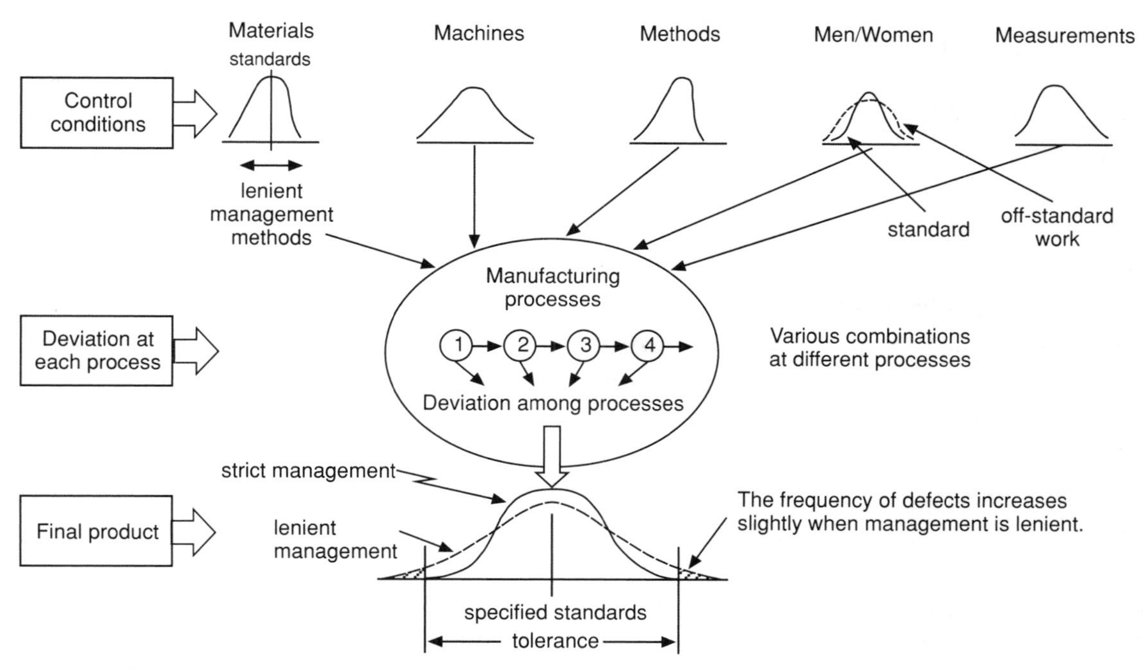

Figure 5-1. The Quality 5Ms and the Approach to Quality Deviation

One day, an improvement group came to observe the process. This group included a very capable IE engineer, RK, who analyzed the cutting machine operations.

RK took an approach that was a bit different from those taken by others before him. He closely observed a veteran operator handling the cutting machine and began analyzing the cutting machine operations. At each step, however, he asked the operator, "Why are you doing that?" By continually asking "why?" in this way, RK concluded that the root problem was friction in the bearings. This friction caused the cutting machine to act a bit irregularly; the minor adjustments the veteran operator made always aimed at minimizing the effects of this irregularity.

This established the focus for improving the cutting-machine process, which was to eliminate the need for minor adjustments. Because RK had discovered the principle behind the machine's irregular operation, he was able to determine the precise adjustment needed to control it.

Analyzing the root cause of the problem, the improvement group found that bits of metal in the lubricating oil created the bearing friction. They solved this problem by installing two preventive devices: an oil filter and a magnet to attract and catch metal fragments in the lubricating oil. As a result, the cutting machine defect rate dropped to zero — even when run by novice operators! This is an example of standardization that clarifies technical principles related to the cause of defects so that all operators can build quality into products.

The most important lesson from this example warns against falling prey to the common belief that some machines are "tricky" and require the mysterious skills of a veteran operator. By understanding the machine's structure and the principles by which the machine builds quality into products, you can standardize operations to enable all operators to use the machine successfully.

This example also illustrates the relation between irregular operation of equipment and defects. In this case, the technical principle involved friction in the bearings; this principle led to the standardization that solved the problem. When incorporating such principles into standards, it is helpful to use illustrations or causal-factor diagrams. Figure 5-2 shows a causal-factor diagram for describing defects caused by foreign matter in painting equipment.

Another chart corresponds to the "why?" approach used in the preceding example — the cause-and-effect diagram (also known as a fishbone or Ishikawa diagram). Figure 5-3 shows an example of this diagram. Some brief instructions for creating these diagrams follow.

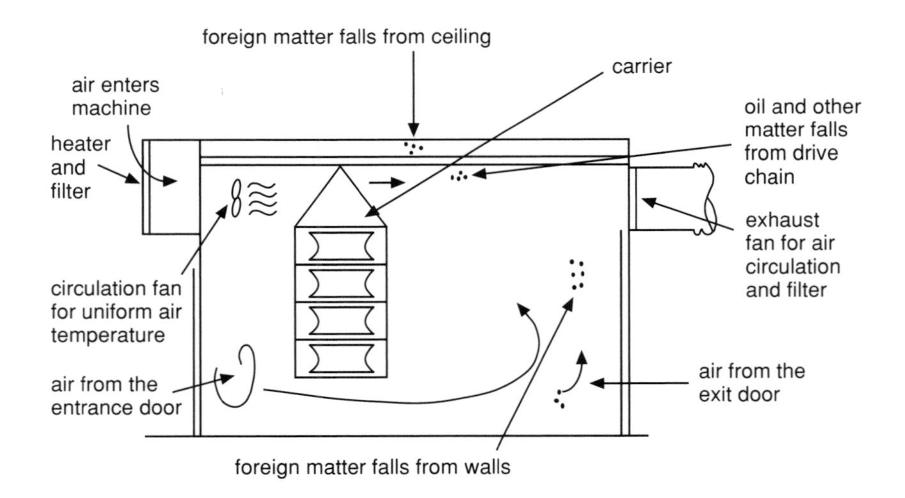

Figure 5-2. Causal Factor Diagram (Painting Equipment Defects)

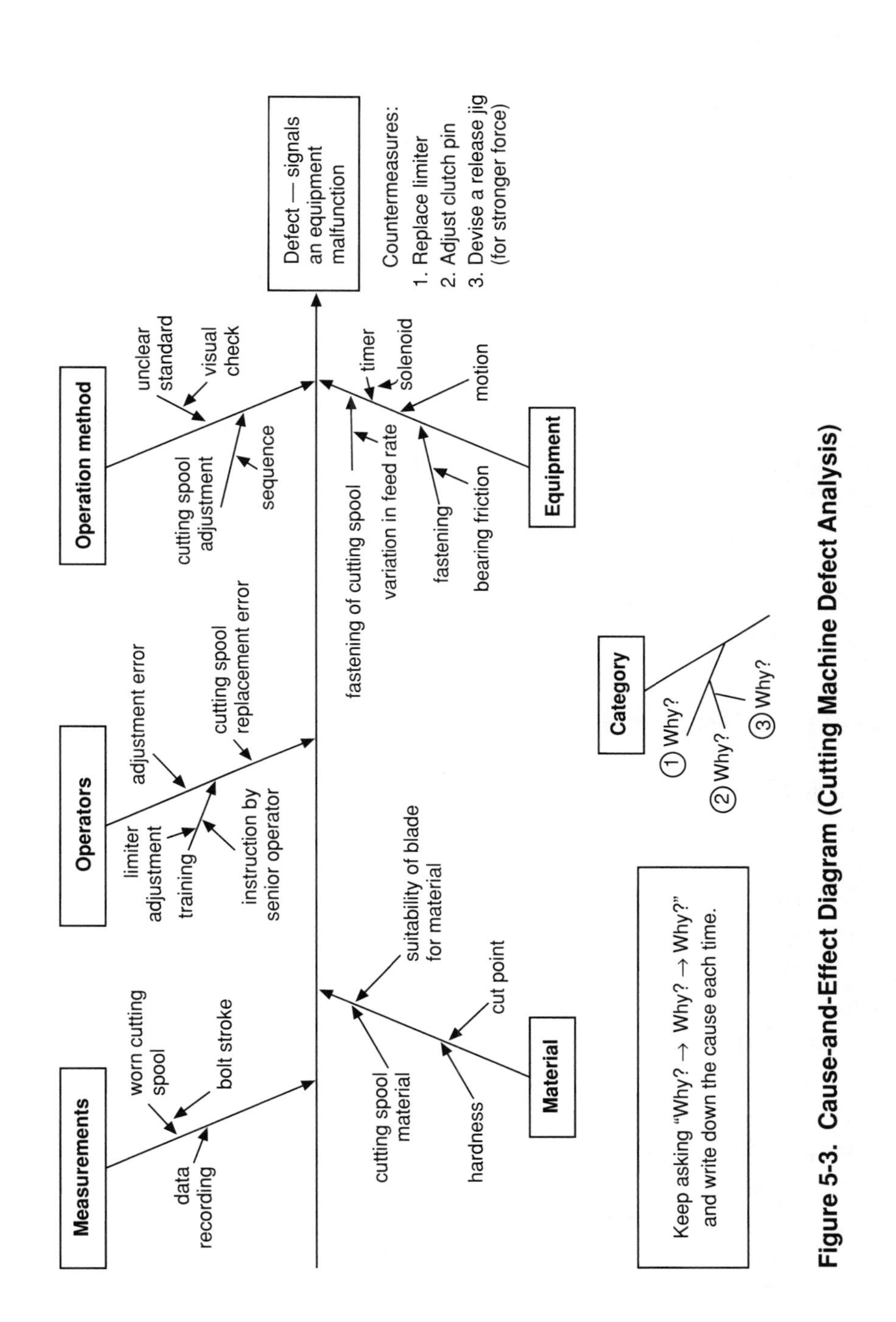

Figure 5-3. Cause-and-Effect Diagram (Cutting Machine Defect Analysis)

How to Create a Causal-Factor Diagram

1. Draw a simple sketch of the mechanism.
2. Label the items that cause defects.
3. Identify and point out in the drawing the cause for a specific defect (in Figure 5-2, the source is foreign matter in the room).
4. With the input of everyone involved, describe the problem caused by each factor.
5. Verify the diagram by checking it against actual work-site conditions. The diagram does not have much significance until its hypotheses are backed up by work-site verification. In the example shown in Figure 5-2, the diagram can be verified by examining the kinds of foreign matter found in defective product samples from the work site. Foreign matter samples can be collected by applying adhesive tape to areas of the work site that are the sources of foreign matter (this is an application of the 5W1H method).
6. After work-site verification, revise the diagram to include any causal factors previously omitted.

How to Create a Cause-and-Effect Diagram

1. On a blank diagram, enter the 5M categories as labels for the main "ribs," then write the causes of problem that fall under each of these main areas.
2. Go to the work site to observe the actual situation. Using the 5M categories, ask "why" to find related causes and effects.
3. Take corrective action against any problems for which you can identify causes.
4. If you cannot find the cause for a certain problem, move on to the next problem, following the same method. As you search for causes and test different effects, your diagram will fill up. It is possible to make a fairly complete list of causes and effects in an office and then verify them at the work site, but it is better to draw the diagram at the work site while you do the verification. Solving problems is the goal and drawing the diagram is just a means to that end. Therefore, it does not matter whether the diagram is neat or well-organized — it is a good diagram if it helps solve problems.
5. Take corrective action against remaining problems.

6. Review the 5Ms and the list of causes and effects, and add standardization for key control points you identify.

In the example shown in Figure 5-3, the improvement group followed up the improvements noted by implementing 5S* (industrial housekeeping), making equipment maintenance standards, and standardizing operation procedures.

DETERMINING CUSTOMER NEEDS TO ACHIEVE ZERO COMPLAINTS

Second Factor for Quality Improvement: Understanding Customer Needs and Use

No matter what type of work you do, knowing what products customers want makes a critical difference. This is particularly true in factory standardization.

Consider the example of factory H, which processes steel rods. The dimensional specification for the rods is noted on the standard sheet as $10^{\pm 0.01\phi}$. Sometimes the factory received complaints about defective rods. A representative of the factory met with the disgruntled customer to find out exactly why the rods failed to meet expectations.

Investigation revealed that a numerically controlled (NC) machine at factory H was cutting the steel rods to 9.8^{ϕ}, due to a problem in the NC machine's auto-feed device that caused the ends of the steel rods to get caught.

To resolve this problem, factory H began using steel rods with tapered ends. This not only restored dimensional accuracy to the specified range and prevented further customer complaints, but it also increased the NC machine's processing speed and eliminated small line stops that the untapered rod ends had caused. The factory people and the customers were equally delighted with these improvements.

Quality (required quality) includes true characteristics and substitute characteristics. In this case, the substitute characteristic was the specification of 10 ± 0.01, a value determined with the client that falls within the range of values and does not pose a production obstacle. The true quality characteristic is

---

* 5S refers to five Japanese words that describe the basic principles of industrial housekeeping: *seiri* (straightening and organizing), *seiton* (arranging), *seiso* (cleaning), *seiketsu* (cleanliness), and *shitsuke* (discipline or adherence). These principles form the basic foundation for many improvement efforts. — Ed.

the countermeasure (rod-end tapering), which enabled the auto-feed device to operate without obstruction and also enabled the NC machine to operate at a faster processing speed.

If the manufacturer can understand a product from the customer's perspective, it can confidently proceed to turn out good products. This is the first step in making products that satisfy customer needs.

Figures 5-4 and 5-5 illustrate two examples of in-process quality standards similar to the tapered rod example. Figure 5-4 shows an example of quality standards in a painting process, and Figure 5-5 shows a standard for the dimensions of cutting tools. Both involve visual quality standards. They serve as good models of standards because they incorporate checkpoints for required product characteristics and are written in an easy-to-understand style.

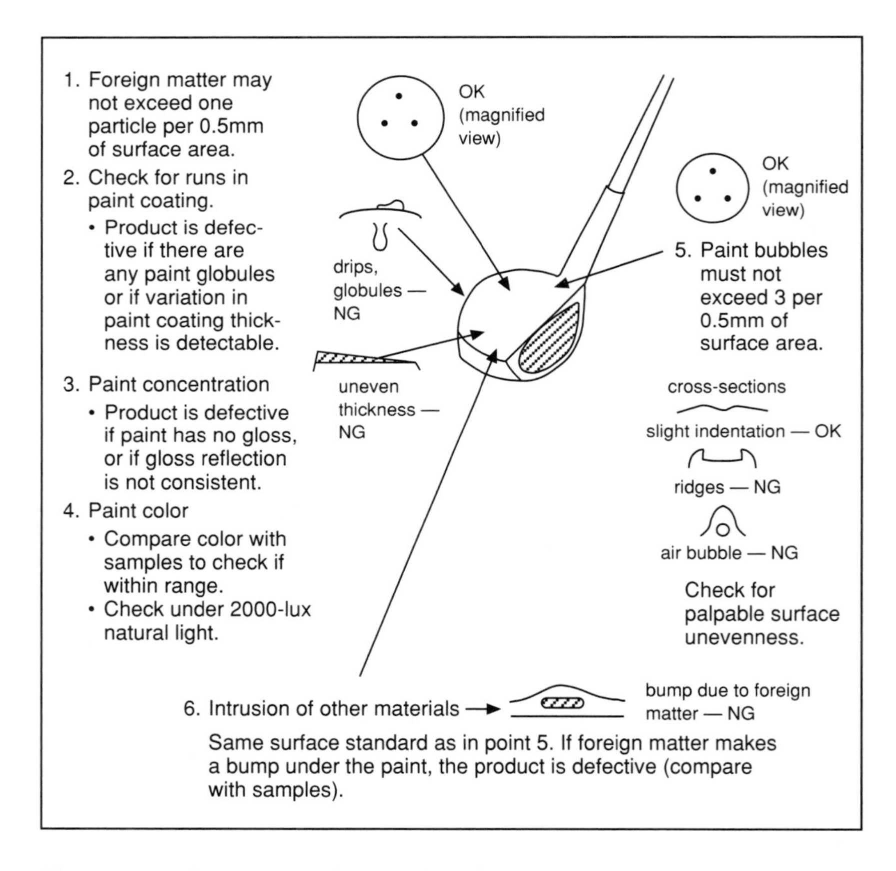

Figure 5-4. In-process Inspection Standard (Quality Standard for Painting Process)

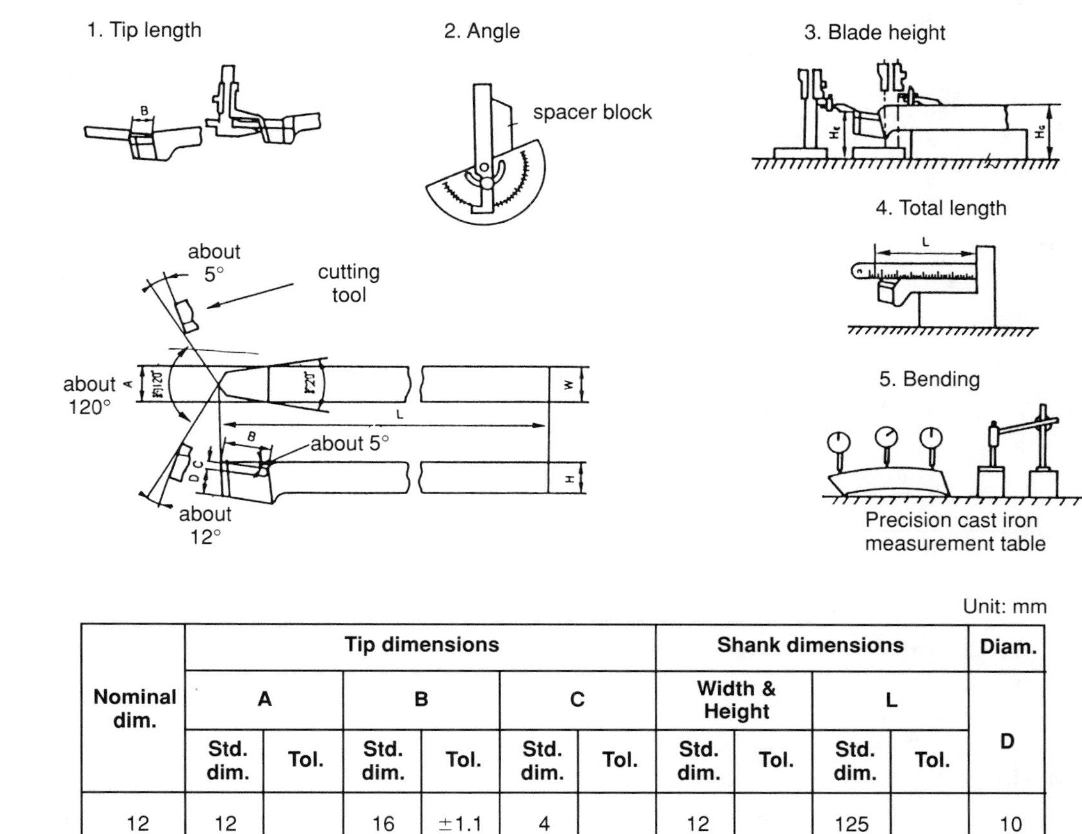

Unit: mm

Nominal dim.	Tip dimensions						Shank dimensions				Diam.
	A		B		C		Width & Height		L		D
	Std. dim.	Tol.	Std. dim.	Tol.	Std. dim.	Tol.	Std. dim.	Tol.	Std. dim.	Tol.	
12	12	±1.1	16	±1.1	4	±0.75	12	±0.5	125	±4	10
16	16		20	±1.3	5		16		140		12
20	20	±1.3	25		6		20	±0.8	160	±5	16
25	25		32	±1.6	8	±0.9	25		200		20
32	32	±1.6	40		10		32	±1.0	250	±6	25

Note: The blade must not be higher than the shank surface.

Source: Japanese Standards Association, *JIS Handbook (JIS Handobukku)* (Tokyo: Japanese Standards Association, 1990).

Figure 5-5. Cutting Tool Checking Standard (Based on JIS Standard)

CREATING DISTINCTIVE QUALITY

Most products that are manufactured, marketed, and used by consumers must compete with similar products made by other manufacturers. Companies know that their products represent them in the marketplace. To compete successfully with rival brands, companies must find ways to set themselves apart from the competition by building distinctive quality into their products.

How is distinctive quality created, and what kinds of quality distinctions are most important? To answer these questions, you must understand quality appraisal factors, and you must also be able to evaluate production operations from the standpoint of the 5Ms — the basic factors that determine quality.

Quality Appraisal Factors

Quality appraisal factors mean two things:

- *Factors that distinguish the product from its competition.* The production managers plan and carry out policies to build distinctive quality into products.
- *Factors that reinforce the product's marketability.* The product planning engineers identify and emphasize design factors that enhance the product's marketability and make the most of its distinctive features.

Figure 5-6 lists several specific quality appraisal factors and organizes them into a matrix to make their interrelationships easier to understand. Operators should keep these factors in mind as checkpoints for standardization to meet customer needs.

Quality Factors within the "5M" Categories

Figure 5-1 presents the 5M categories of factors that determine quality. Although these categories will be discussed in more detail later, a brief discussion here will introduce some basic considerations.

Men and Women (the Work Force)

One of the major quality factors in this category is enthusiasm. It is no exaggeration to say that the "can-do" spirit is a basic determinant of product quality. Key quality elements of this enthusiasm in the work force include:

- A clear knowledge of the purpose of one's own work
- A sense of the importance of one's work
- An in-depth mastery of correct work methods

Methods

Work methods are another basic determinant of quality. These methods are described in process standards, which will be discussed in Chapter 7.

Machines (Equipment, etc.)

This category is best thought of in terms of using equipment to build quality into products; this category will be discussed in Chapter 6.

In ancient times, people believed that gods helped control the fire and temperature in foundry furnaces and cupolas. Today, this "divine" knowledge of temperature control has been analyzed and quantified by researchers and incorporated into software to enable computerized control of such equipment.

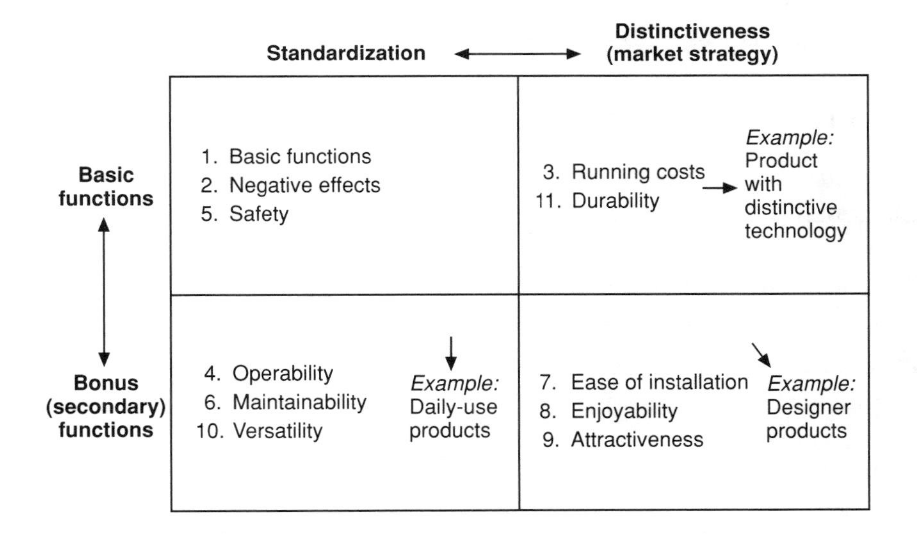

Figure 5-6. Product Quality Appraisal Factors and Their Significance

Materials

It is common knowledge that the quality of the materials has a big impact on the quality of the product. Accordingly, most companies have a full complement of standards for checking the quality of materials.

However, mysterious quality problems still crop up now and then. For example, companies sometimes find that although materials have passed a thorough inspection, they still create quality problems when used in products. Often the culprit is an additive in the material. Perhaps the additive in previous batches was more uniformly dissolved into the material than in the defective batch. Some additives cannot be trusted to react with the base material in the same way. It is not always possible to detect such problems by inspecting only the separate ingredients.

This phenomenon tends to occur in such items as welding rods, solvents, and deoxidizers. Manufacturers should be especially alert for it when switching to new materials suppliers.

Measurement

Products must always be manufactured in accordance with technical standards. Measurement standards are one type of technical standard; they are critically important in determining product quality.

Types of measurement standards include the following:

- Passage inspection, counting, etc.
- Dimension and position checks
- Safety and alarm checks
- Temperature, humidity, and viscosity checks
- Flow rate, voltage, and current checks
- Inspections for defects and loose parts
- Ingredient measurements
- Level and marker checks
- Color, identification mark, and light intensity checks

Many of these measurements and inspections can be automated. Measurement-related quality can be controlled not only by sensors but also through mistake-proofing methods, which are an important means of standardization. Figure 5-7 shows some mistake-proofing improvements.

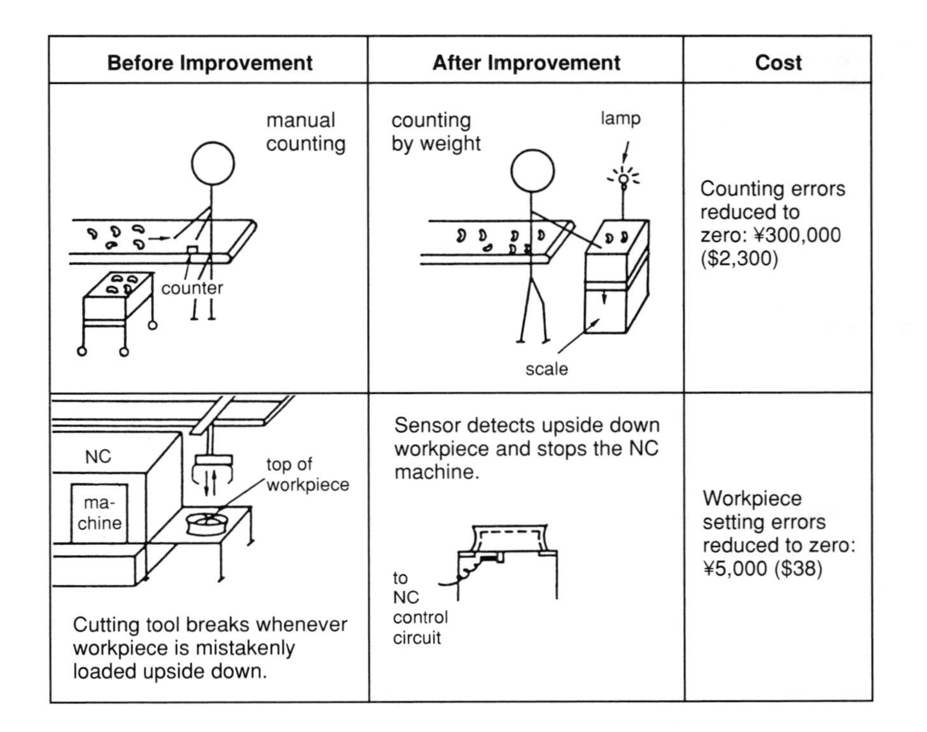

Figure 5-7. Examples of Mistake-Proofing Improvement

STANDARDIZATION OF PROBLEM ANALYSIS

At this point the discussion moves from quality improvement concepts to practical methods for implementing these concepts. The steps for carrying out standardization are P for Plan (i.e., planning and adopting improvement proposals), followed by D for Do (implementing the improvement plan using the most effective methods or standards), then C for Check (i.e., check the effects of the improvement at the work site). The cycle begins again with P for Plan (plan further improvements to correct any remaining abnormalities). This P-D-C cycle is a basic concept for systematic quality control. The following sections examine specific standardization methods that can be applied during cycle.

The Importance of Maintaining Standards

The upper left part of Figure 5-8 shows standards made at a factory to prevent product defects. It was noted that the work site could continue operating

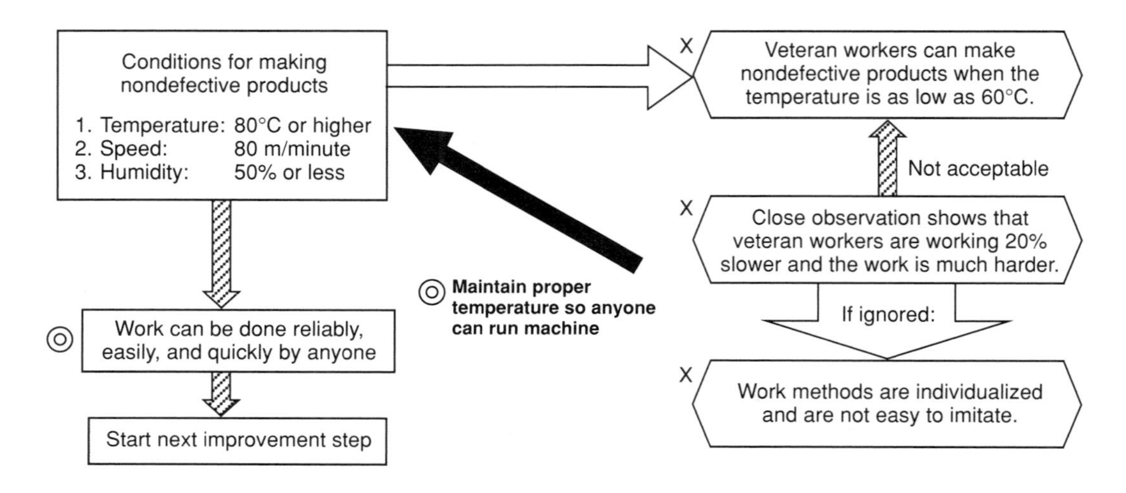

Figure 5-8. Approach for Maintaining Standards (Example)

without producing defects when the temperature was as low as 60° C (this process includes heat treatment). A close inspection revealed, however, that the process operations moved 20 percent slower at that temperature. The speed reduction was the result of extra work the operators performed to compensate for the lower temperature.

This kind of operator response is like a bad habit. If it is ignored or accepted, the operators will learn to deal with off-standard heat treatment temperatures by taking extra steps of some kind. The off-standard work environment creates a work site that newcomers cannot operate as well as veterans who have learned the "tricks" for dealing with varying conditions. Such tricks undermine trust in standards and lead to a work environment that runs on individual styles and roundabout measures to compensate for problems.

If standards are not enforced, operators become less likely to devise better conditions. They are also less likely to wonder about making improvements, and, as a result, the work site is less likely to keep pace with recent technological advances.

These and other facts underscore the importance of standards. You cannot move toward the next stage of improvement until you have established steady maintenance of the standards already in place.

Basic Steps in Standardization

One lesson of the speed reduction example proves the mistake in thinking, "We are maintaining the standards, so we have no defect problems." Unless quality standards are very high, there is no reason to believe that further quality improvement is unnecessary.

There are many possible reasons for ongoing defects. Perhaps the current standards contain errors, or perhaps some defect causes have not yet been identified, or maybe the current standards are hard to follow and need improvement. That is why problem-consciousness and skill in analyzing problems are "musts" for effective standardization.

Another "must" is the QCD approach — assertively analyzing and solving the work site's quality, cost, and delivery problems through equipment and other improvements, and creating a higher level of standards that integrate the QCD factors. Even when you manage to establish a very high level of standards, the fast pace of technological progress soon makes these standards obsolete. You must realize that standards are always subject to improvement. This is expressed in the popular QC maxim: Improvement is endless.

Figure 5-9 describes the basic approach to standardization, from analyzing problems to making new standards.

CASE STUDIES: HOW TO ANALYZE PROBLEMS

Problem-consciousness is the starting point for standardization. The concept of problem-consciousness is easy enough to understand, but the problems themselves are not always easy to find, even when being problem-conscious. Use the checklist shown in Figure 5-10 as an aid for finding problems.

When applying QC methods to find problems, it is especially important to use the 5W1H method to gain an in-depth understanding of the facts, repeatedly asking "Why?" to discover the root causes of problems (see Figure 5-11).

A willingness to go to the work site to discover the facts and an eye for wasteful operations are basic requirements for improving the level of standards in the future. The following case studies illustrate the proper approach to analyzing problems.

Case 1: Cracks in Cast Items

A foundry was plagued by defective goods, mainly cracked cast items. Although the defect rate was only 0.1 percent, the large volume of items produced created an unacceptable number of bad products.

The improvement group set up a display showing the formula for calculating the defect rate, the number of defective items, and the monetary value of

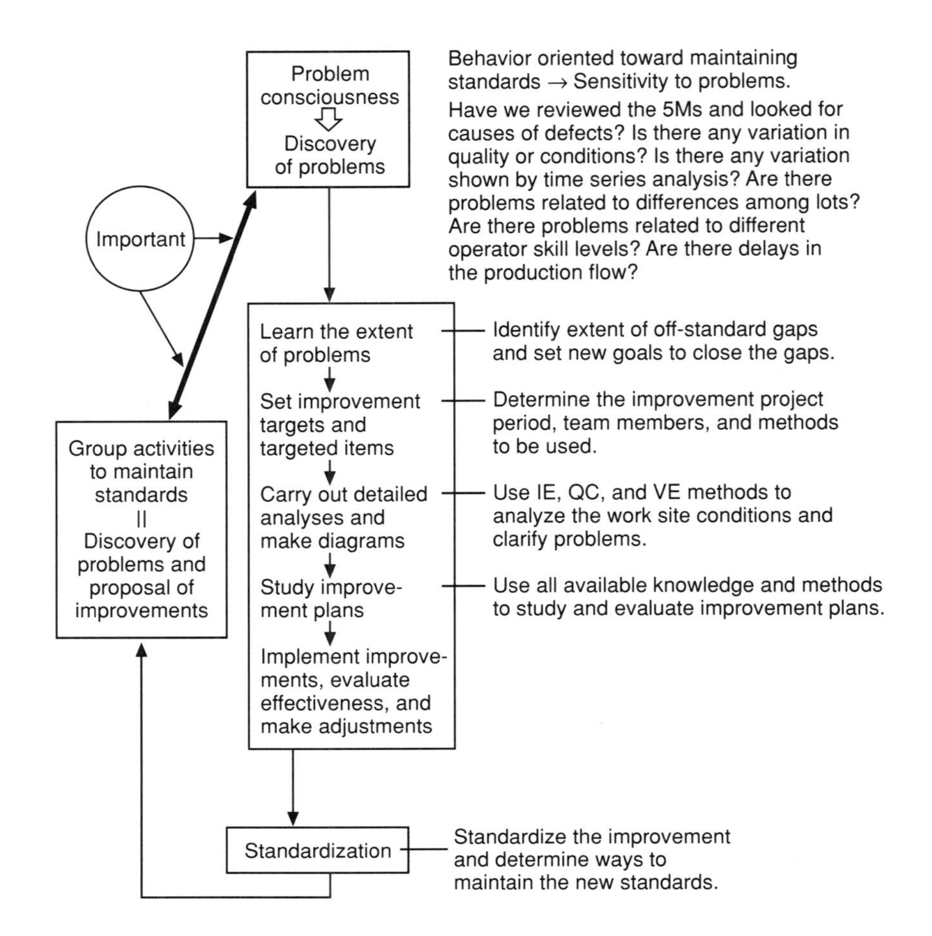

Figure 5-9. Steps from Problem Analysis to Standardization

the defect loss. They also painted the "defects" bin bright red and placed it conspicuously along the walkway in the plant. At the time, the plant management had just begun a 5S campaign. When setting up the defect display, the managers also posted a statement saying that "Defect loss amounts to ¥500,000 ($3,850 at ¥129.87/$1) a month. Eliminating this loss would save the equivalent of one month's pay." This statement drew particular attention because it coincided with the beginning of Japan's annual "spring labor offensive" of wage negotiations; companies tend to be more generous with wage increases when money-saving efforts work.

The improvement group began by going through an SN-type problem-finding checklist (shown in Figure 5-10), asking the 5W1H questions to trace the source of the defects. They also implemented full-lot inspection.

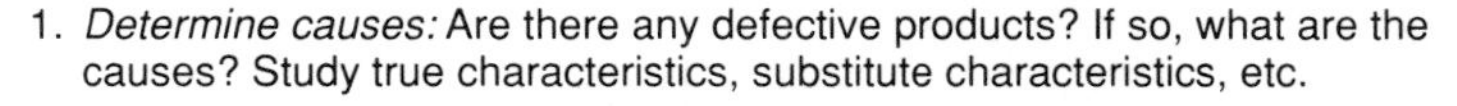

1. *Determine causes:* Are there any defective products? If so, what are the causes? Study true characteristics, substitute characteristics, etc.
2. *Investigate variation:* Are the quality mean values appropriate? Is the amount of variation appropriate?
3. *Classify:* Are there differences among groups? Classify by furnace, lot, job post, etc.
4. *Organize causes and effects:* If several problems are found, which are the most important? Learn to quickly draw up cause-and-effect lists that include at least 30 items.
5. *Analyze quantitatively:* How can the problems be described in quantitative terms? Which is the most frequent problem? Use Pareto analysis.
6. *Study variations in time series:* Taking current standards as the base, have you done a time series analysis to look for variations that may indicate problems?
7. *Understand process links:* Does this process cause time-consuming problems at other processes? Does it create some other kind of problem for them? Does information flow smoothly among processes?

Figure 5-10. SN Problem-Finding Checklist (for Quality Control)

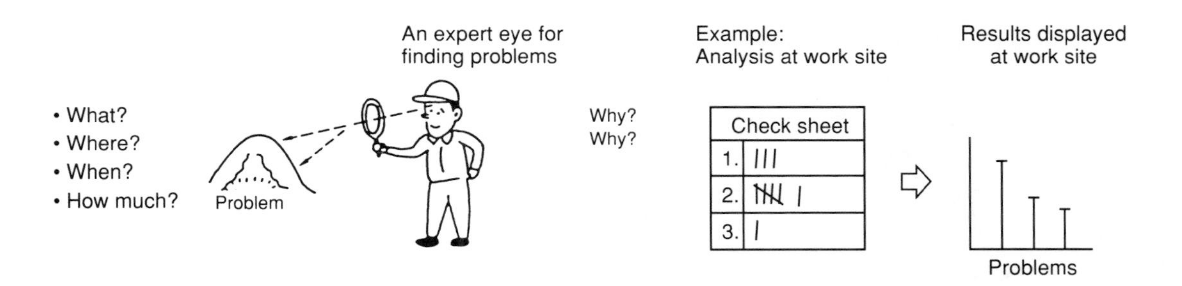

Figure 5-11. Finding the Facts at the Work Site

Their investigations and observations revealed that after casting, when the conveyor dropped finished products into baskets for heat treatment, some of the products sustained cracks from the impact. The group devised various ways to soften the impact, such as lining the baskets with cushioning material, reducing the gap between the conveyor and the baskets, and having the products fall through a rubber net before reaching the basket. After these improvements, they found they had reduced the defect rate to the level of parts per million.

Activities based on observing current conditions at the work site, which in this case included full-lot inspections and detailed process analyses, may require a lot of labor-hours. Many factories have problems that require such a thorough approach. However difficult and time-consuming it may be, a solid understanding of the facts provides a very effective means for solving deep-rooted problems.

Case 2: Poor Die Closures in Casting Equipment

Precision in die closures is especially important for press processes that use metal dies closed under hydraulic power. The hydraulic pressure in such processes is usually several hundred tons. The factory in question began having problems when some of the molten material leaked from the dies, causing poor separation of the cast items from the dies.

The managers were surprised that such problems arose, since the casting equipment was still new. The improvement group checked the dies and found nothing wrong. They also checked the cylinders — still nothing wrong. The problem was that pressure was being applied too slowly to the dies.

They applied time-series analysis to come up with a list of problems and closely studied the variations in the time series to locate the problems. They found that after production start-up, the oil temperature rises as it passes through the cylinder that presses the die closed. When the oil temperature reaches a certain range, the hydraulic pressure starts to decrease. This is similar to the phenomenon of bubbles appearing in automobile brake fluid when the fluid overheats during prolonged downhill braking.

The improvement team measured the oil temperature and the flow rate and confirmed that the oil was overheating. They drew diagrams to illustrate the problem and to trace the path of the oil flow. This led them to suspect that something was wrong with the filter. They cleaned the filter, and suddenly the overheating problem disappeared.

Unfortunately, the same problem recurred three months later. The team checked the filter and found it had already become dirty, much sooner than expected. This led them to check the oil itself, and they found the oil quality had deteriorated. They changed the oil, flushed out the hydraulic system, and installed a new filter. As a preventive measure they decided to implement a filter replacement every three months; however, the filter was not dirty after the next three months, and the casting machine has been operating smoothly ever since.

Some improvement groups take a "shotgun" approach when they have several problems: They aim at solving all of them and in the process solve at least one or two. But this approach wastes a lot of time and effort in struggling with problems that do not get solved. It is wiser to take a "sharpshooter" approach that focuses on a single problem and applies the most suitable method for solving it. This approach is especially effective when the improvement group benefits from the experience-based skills of one or more veteran workers.

Case 3: Broken Belt in NC Machining Center

NC machining centers use timing belts to precisely control the operation of cutting tools. At this factory, the NC machining center's timing belt occasionally broke, causing defects. Overloads sometimes occurred, causing the workpiece to come loose and damage the machining center. The factory managers were very concerned with the resulting stoppages.

They called in a technician from the equipment manufacturer to replace the timing belt, a procedure that took six to seven hours. They even signed a regular maintenance contract with the manufacturer to help avoid further timing belt problems, but the problems still happened. At one point, four of the ten production lines had NC machining centers with broken timing belts.

Realizing that relying on the manufacturer was not working, the factory managers formed a project team to deal with the timing belt problem.

The team first went to the work site to investigate the facts. They found that the boxes that contained the timing belts also contained some cutting chips. The boxes were also poorly fastened and sealed, allowing the intrusion of chips and other foreign matter. The team devised a new timing belt replacement method using different tools, which reduced the belt replacement time to just one hour. They standardized some of the other tools and jigs to reduce variety and devised new parts-replacement procedures for these standardized items. They also repaired the timing belt boxes to fasten them more securely and seal them more tightly. However, these improvement measures were only temporary solutions.

The team's next step was to repeatedly ask "Why?" in pursuit of the root causes. Although they suspected that the belt material could be improved, they also considered such factors as the presence of cutting chips in the timing belt boxes, the belt boxes' loose fitting, and the fact that the belts always broke in the same place. They repeatedly asked "Why?" to find the root cause of each situation.

One of the factors explored was that chips were getting behind the platform where the cutting tool operates. This was cleaned out occasionally, but the chips did not always accumulate in the same places, and the cleaning was not done on a frequent or routine basis. Consequently, the chips accumulated to the point where they could press against the timing belt box enough to cause the box to loosen.

After making a video recording to analyze the directions in which the chips were cast from the cutting tool, the team installed a protective cover (designed for easy removal and replacement) and standardized the tools, methods, and schedule for chip removal.

Together, these various measures prevented further timing belt breakage. This case study illustrates how, along with advances in cutting technology, controlling cutting chips is important for improving cutting processes.

STANDARDIZATION OF KNOW-HOW

The P-D-C cycle of standardization ends each round with C (meaning "check" or "see"). Most of the time, nothing will go wrong if you only identify the problems and implement improvements without checking up on them afterward. But doing so entails a risk that defects will recur and labor-hours will be wasted. Operators need know-how to check on a process while doing a good job of handling it. Without such know-how, standardization remains superficial and eventually loses its usefulness.

This section examines a method for identifying know-how control points as well as standardization methods. There are various kinds of know-how, as several case studies show.

Good Know-How and Bad Know-How

What happens when you throw a magazine onto a campfire? Even though the magazine is made of paper, it doesn't burn very well. (Perhaps you've tried this experiment as a child.) Next, you pull the magazine out of the fire, open it, and crumple the pages to allow air space between them, then set it back into the campfire. Now it burns quickly.

Ways to deal with large, dense objects have been a part of human "know-how" for a long time. Once we understand the basic principles of a problem, it becomes much easier to solve.

Consider the problem of evenly heating four flat objects stacked on several shelves. The standard operation is to leave a gap between each shelf. The know-how behind this standard is fairly obvious. If hot air is circulated around the shelves, the space increases the exposed surface area, allowing it to have similar contact will all the objects.

Once you possess such know-how, there are many ways to apply it. The principle of maximizing the exposed surface area can be used for both heating and cooling processes, such as in vibrating a large object underwater to help it cool faster. Likewise, we can use this principle to figure out that spraying a mist (water and air) cools an object more effectively than just pouring water on it.

The results two people achieve when separately performing the same kind of work differs greatly when one person applies know-how principles and the other does not. Knowing such principles and being aware of opportunities to set new standards are equally important; however, there is bad know-how as well as good know-how. Bad know-how includes the tricks and subtle methods used for dealing with nonstandard situations rather than preventing their occurrence. When you analyze problems in the key 5M categories — men/women (people), methods (operations), materials, machines (equipment), and measurement — you can recognize when something is wrong with the situation or with the approach for dealing with it.

The late manufacturing expert Shigeo Shingo sometimes used a know-how example that involved wearing two pairs of socks to prevent blisters from tight golf shoes. He observed which surface interface had the lowest coefficient of friction: between the shoe and the outer sock, between the inner sock and the outer sock, or between the foot and the inner sock. He found that the second interface had the lowest friction coefficient, and concluded that wearing two pairs of socks was the best way to avoid friction and blisters.

Dr. Shingo once applied this principle to help improve the process of mounting the workpiece on a milling machine. As the workpiece was bolted down, the turning of the bolt caused it to shift slightly out of position. Veteran operators knew this and tapped the workpiece to keep it in place while turning the bolt. Referring to how two pairs of socks reduce friction on the foot, Dr. Shingo suggested they use two washers for the bolt, with lubricant between the washers, then position the workpiece and give the bolt a quick final turn. The workpiece stayed in place perfectly when bolted down this way.*

The painstaking tapping method practiced by the veteran operators was one type of know-how. However, it was not a method that guaranteed success,

---

* For another example of Dr. Shingo's application of this principle, see Shigeo Shingo, *A Revolution in Manufacturing: The SMED System* (Cambridge, Mass.: Productivity Press, 1985), 71-72.

nor was it easily taught to other operators. Dr. Shingo's response to such a situation was, "Not know-how, but know-why." In other words, instead of simply learning a method for working around a problem, you should learn *why* the problem exists in the first place; then you can solve the problem instead of working around it. This is an important point. Much of the know-how practiced in factories today is only a laborious and time-consuming way of working around problems instead of solving them.

Making Standards That Emphasize Know-How

The following example uses a series of steps that includes the problem-finding method, P-D-C cycle, and the standardization method described above. This example, a case study in soldering operations, is illustrated in Figure 5-12.

Step 1: List the Problem Points

A veteran operator watches a new operator perform the soldering operations and makes a list of the problem points observed. Later, the veteran operator ranks the items on the list (on a scale of 0 to 5 with 5 as top priority).

Priority	
5	1. Some connection points are weak.
4	2. The soldering iron burned some wire.
4	3. Most of the solder adheres to the plug instead of the socket.
4	4. The plug moves during soldering, which makes it harder to solder.
3	5. The solder does not flow well on the soldering iron tip, and grinding down the tip is difficult.
4	6. The soldering heat sometimes burns or cracks the plastic.

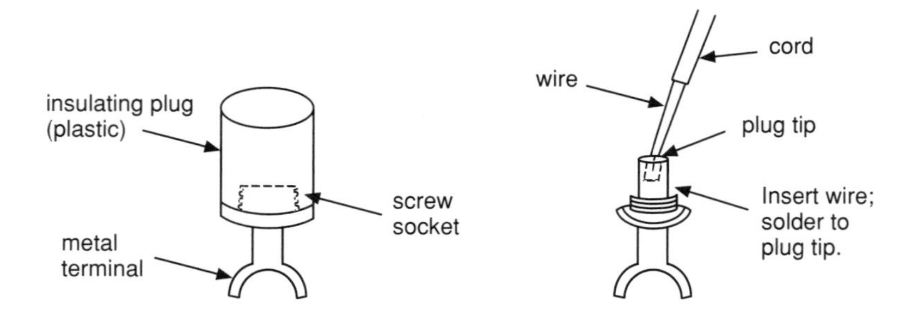

Figure 5-12. Case Study: Soldering Operations

Step 2: Study Countermeasures for the Listed Problems

The veteran operator and a technician look into countermeasures for the listed problems. Their standardization ideas include the following:

1. Ascertain the current technological level and set a new level.
2. Check the operator's skill level and establish a training system.
3. Develop a process for evaluating the effects of improvements (such as to determine whether the improvement is worth the investment costs and labor-hour costs involved).

Figure 5-13 shows a chart for evaluating improvement levels from the standpoints of technology, operators, and cost effectiveness. Figure 5-14 shows a problem-analysis table that can be used to establish improvement measures involving these three dimensions.

Step 3: Clarify the Manufacturing Technologies

If possible, this step should be done with the help of a technician and/or a veteran operator. In fact, all standardization is best done with some expert assistance.

Standpoint	A: Current level	B: Minor improvement	C: Major improvement
T: Technical	Slight improvement in standardization of manual operations and jigs and tools.	Not technically difficult; use successful technologies developed by other departments.	Apply new technologies in safety, quality assurance, areas supported by industrial policy, and new key technologies.
O: Operator	Hire temporary workers and subcontractors to fill skill gaps.	Train operators in multiprocess handling, NC machine operation, etc.	Train operators to use computers and maintain factory automation systems.
C: Cost effectiveness	Raise efficiency of operations; can solve current problems at a cost of about ¥100,000 ($770) per person, resulting in productivity increase of up to 10 percent.	Can raise productivity 20 to 30 percent by reducing the defect rate and breakdown rate, at a cost of ¥500,000 to ¥1,000,000 ($3,850 to $7,700) per person.	Commit heavy investments without regard for short-term return on investment.

Figure 5-13. Evaluation Chart for Improvement Level of Countermeasures

Problem Point	Relevant Facts	Countermeasure Level Evaluation			Improvement Measures
		Technology	Operator	Cost Effectiveness	
1. Some connection points are weak.	1-1. Poor solder flow — solder too slow for effective fluxing.	A ↑ Standardize using a timer	A	A ↑ Develop specialized jig	⇒ 1. Make specialized jig 2. Make training manual 3. Make process standards
	1-2. Varying methods used to check the solder connection. 1-3. No operator understanding of timing and methods for solder cooling.	A ↑ Set a mandatory cooling period	A ↑ Solder cooling: make an easier-to-use jig	A ↑	⇒ 1. Make a jig to enforce solder cooling standard after soldering 2. Make process standards 3. Make technical standards
2. Heat burns wires	2-1. No standard length for plastic section.	A Train operators in correct use of stripper	A	A	

Figure 5-14. Problem Analysis Table

The key point at this step is to ask "why." For example, why is the operator doing the work that way? Operators who understand the "why" behind the "how" find standards easier to maintain.

They also tend to find their work more interesting and are therefore more motivated to do a good job. When they understand the principles behind what they do, they are also better able to devise improved work methods and otherwise improve their work environment.

One way to teach operators about manufacturing technologies is to create work-site displays based on the technical (manufacturing) standards, perhaps including workpiece samples as further illustration (see Figure 5-15).

The technical standard display helps the operators clearly understand the physical characteristics and mechanisms of the operations they are performing, which in the case of soldering include the temperatures at which solder melts and solidifies, the principles of solder connections, that solder connections are weak until the solder is fully cooled, the importance of using flux for removing oxidation layers before soldering and the consequences of not using it, and the need to have the solder connection and the parent material at about the same temperature during soldering.

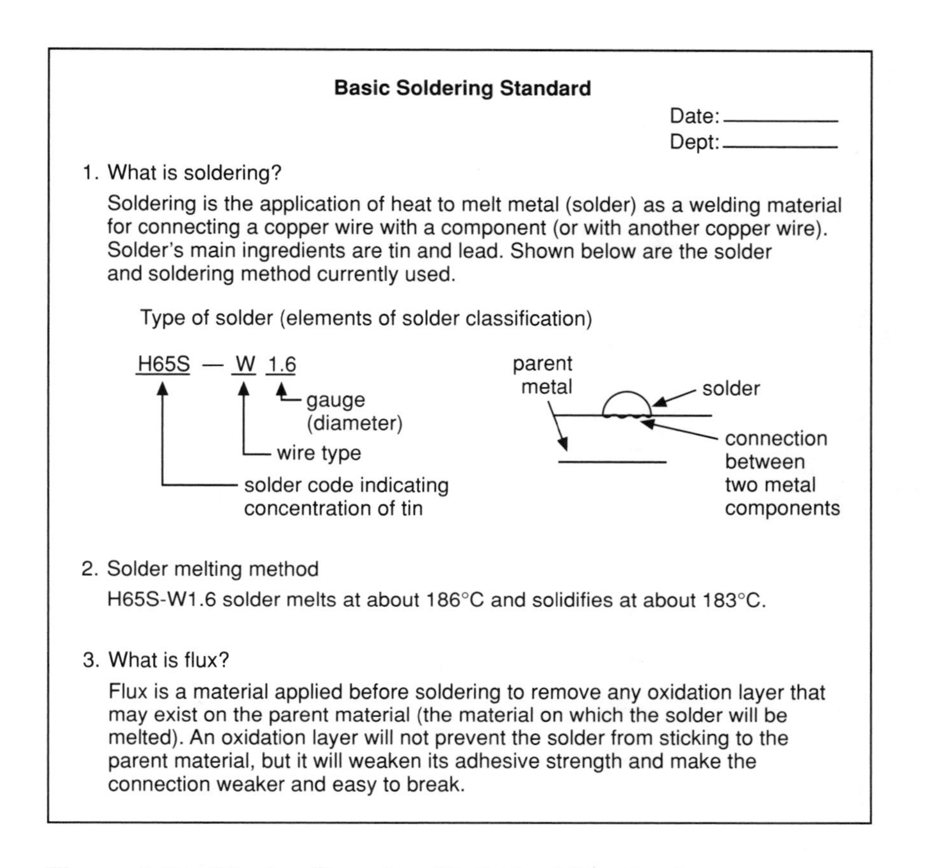

Basic Soldering Standard

Date: ______
Dept: ______

1. What is soldering?

   Soldering is the application of heat to melt metal (solder) as a welding material for connecting a copper wire with a component (or with another copper wire). Solder's main ingredients are tin and lead. Shown below are the solder and soldering method currently used.

   Type of solder (elements of solder classification)

2. Solder melting method

   H65S-W1.6 solder melts at about 186°C and solidifies at about 183°C.

3. What is flux?

   Flux is a material applied before soldering to remove any oxidation layer that may exist on the parent material (the material on which the solder will be melted). An oxidation layer will not prevent the solder from sticking to the parent material, but it will weaken its adhesive strength and make the connection weaker and easy to break.

Figure 5-15. Display Based on Technical Standard (Technical Points for Soldering)

When using new operation standards, check to make sure the operators adhere to the principles of the standards. It should be considered a quality control "must" to instruct operators in the technical standards relevant to their work.

Step 4: Choose the Best Improvement Plan

Several possible improvement plans could be adopted for each problem. The question is how to choose the best plan. Figure 5-16 shows a sheet that lists ideas for improvements, along with their particular objectives or functions. Figure 5-17 shows an evaluation chart for studying the more promising improvement plans (in this case, Plans A, B, and C) in terms of their technical,

Objective (Function)	Ideas	Plan Illustration
1. Attach flux to parent material	1. Use resinous solder 2. Apply flux by hand 3. Apply molten flux to wire	Plan A: Improvement in manual operation (1-1) + (2-1) + (3-5) 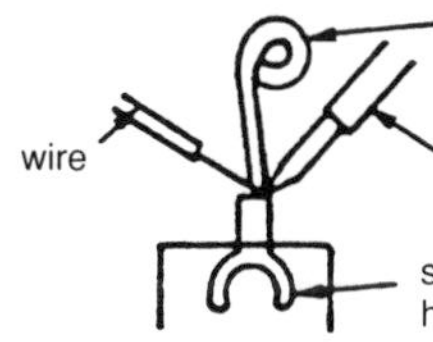 ... resinous solder wire soldering iron stabilizing jig made of heat-resistant material
2. Attach solder to parent material	1. Apply well-soldered iron tip to parent material 2. Apply molten solder to wire 3. Insert solder between wire and parent material; apply heat quickly to melt the solder	Plan B: Preheating method (1-3) + (2-2) + (3-1) ... ① Dip wire in molten flux → ② Dip in molten solder 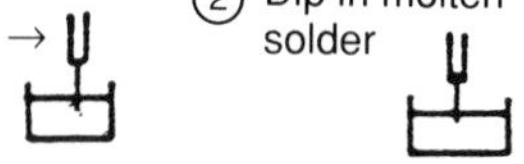
3. Heat parent material	1. Use heater for preheating 2. Heat parent material to solder melting point 3. Preheat flux to melting point	 ③ Apply soldering iron preheating jig Plan C: Spot welding method (1-3) + (2-3) + ...

Note: Numbers in parentheses are combinations of ideas from each of the objectives.

Figure 5-16. Idea Sheet

Idea	Primary Evaluation Criteria			Evaluation
	Technical	Operator-related (including safety)	Economic (including operating costs)	
Plan A: Improvement in manual operation 1. Use resinous solder 2. Apply a large amount of solder to the solder iron tip, then apply solder 3. Use a specialized jig to stabilize the plug	◎ Feasible even under current conditions.	△ Requires veteran skills; training will be needed.	◎ Inexpensive: only solder material needs changing.	○ Try using the specialized jig, give verbal instructions.
Plan B: Preheating method 1. Apply flux beforehand 2. Dip wire into molten solder 3. Set plug into preheating device	◎ Requires only a new jig	◎ Can be done by setting standard sequence	○ Jig improvement can be done inexpensively	◎ Implement Write new standard sheet
Plan C: Spot welding method 1. Apply flux to wire 2. Apply wire to spot welder tip and insert solder for quick melting	○ Requires tests for technical problems (test new materials, etc.)	◎ Easy to do	△ Spot welder investment required: about ¥1 million ($77,000)	△ Postpone

Figure 5-17. Idea Evaluation Chart

operator-related, and economic advantages. This type of chart can be very helpful for comparing plans and selecting the most effective one.

After selecting one of the plans (Plan B, for example), you need to make a more detailed study and perhaps conduct some tests before implementing it.

Step 5: Create Explicit Standards

This step involves introducing the new operation methods at the work site and training the operators to perform them. At this implementation stage, operators must follow the standards precisely. Close adherence to the standards is essential for identifying problems, preventing defects, and making the work go more smoothly and quickly. Standards represent the best techniques a company has for making products. As such, they should be regarded as a valuable part of the company's assets.

Standards and tools are described on the level of operation technology. No matter how good the production machines and other production facilities, if the operation technology is poor, the products will be poor. If work-site machines are difficult to use, you must make improvements to make them easier to use. The front-line operators are chiefly responsible for maintaining a problem-conscious attitude and for spotting hard-to-use tools, jigs, and other equipment.

Figure 5-18 shows an example of a standard operations sheet that takes this use-oriented approach. Since the operations described in this example are rather simple, this sheet includes both the technical standards and the process standards.

Whenever you draw up a new standard sheet, check it over from the QCD perspectives (quality, cost, and delivery, or in this case operation time). You should also do a safety check, using a separate checklist that should include the following types of check items:

1. Looking at quality, have we clarified the process defects and background causes for past quality complaints? Were corrective measures taken? Is there any variation in quality among different operators?
2. As for costs, are there any unnecessary materials or parts?
3. As for operation times, is the sequence of work difficult to understand or follow?

The checklist should also check the problem points identified in Step 1.

When the new standard sheet has been thoroughly checked, it should receive an approval stamp as a "factory standard." It is then ready to be introduced and maintained at the work site.

Step 6: Activities to Maintain Standards

Only when standards are maintained can you expect to make good, reliable products inexpensively and quickly. Figure 5-19 shows Canon's approach to maintaining standards. As mentioned in Figure 5-8, you must maintain the current standards before you can go on to raise standards to a higher level.

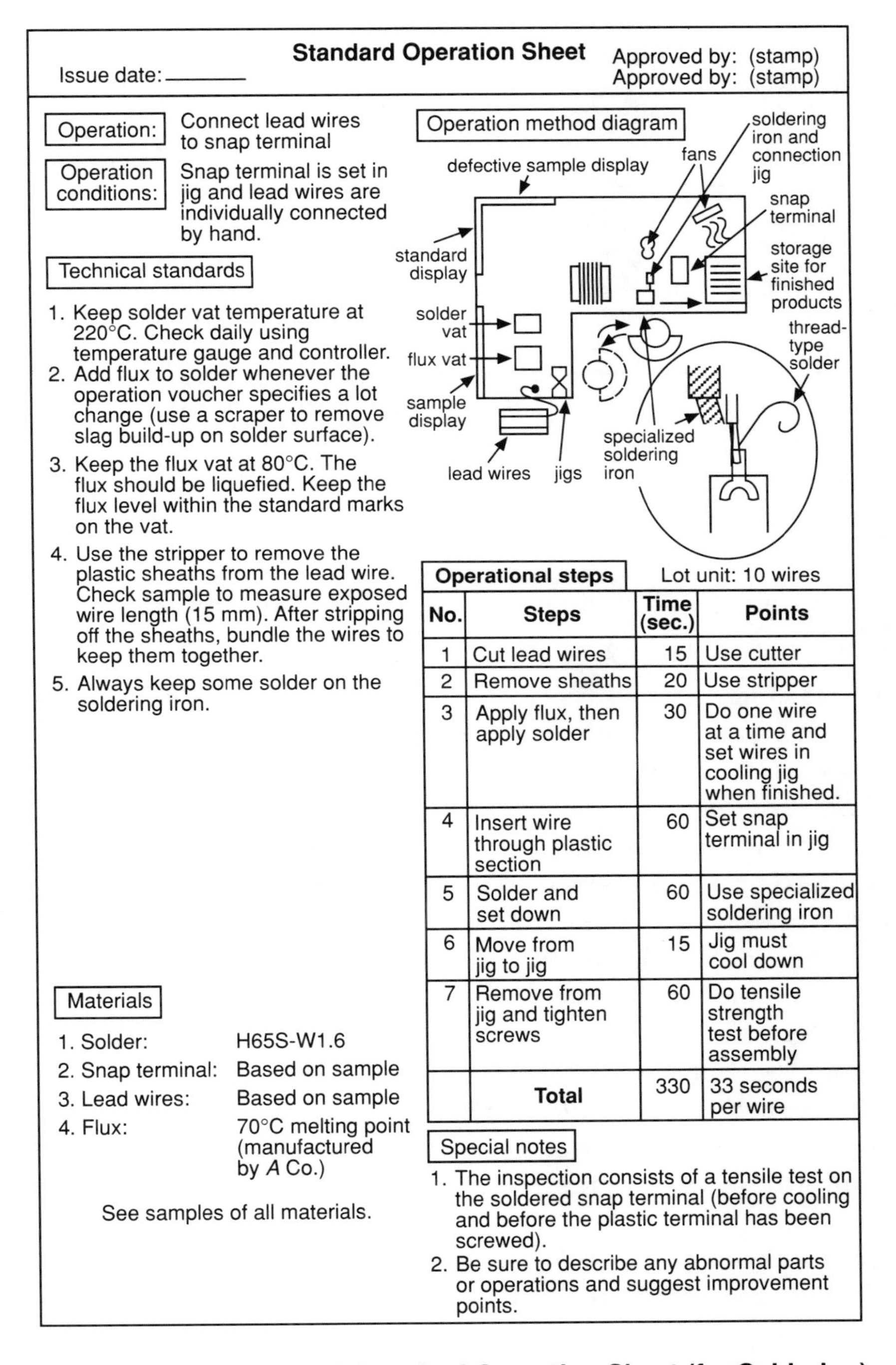

Standard Operation Sheet

Issue date: ______

Approved by: (stamp)
Approved by: (stamp)

Operation: Connect lead wires to snap terminal

Operation conditions: Snap terminal is set in jig and lead wires are individually connected by hand.

Technical standards

1. Keep solder vat temperature at 220°C. Check daily using temperature gauge and controller.
2. Add flux to solder whenever the operation voucher specifies a lot change (use a scraper to remove slag build-up on solder surface).
3. Keep the flux vat at 80°C. The flux should be liquefied. Keep the flux level within the standard marks on the vat.
4. Use the stripper to remove the plastic sheaths from the lead wire. Check sample to measure exposed wire length (15 mm). After stripping off the sheaths, bundle the wires to keep them together.
5. Always keep some solder on the soldering iron.

Operation method diagram

Operational steps Lot unit: 10 wires

No.	Steps	Time (sec.)	Points
1	Cut lead wires	15	Use cutter
2	Remove sheaths	20	Use stripper
3	Apply flux, then apply solder	30	Do one wire at a time and set wires in cooling jig when finished.
4	Insert wire through plastic section	60	Set snap terminal in jig
5	Solder and set down	60	Use specialized soldering iron
6	Move from jig to jig	15	Jig must cool down
7	Remove from jig and tighten screws	60	Do tensile strength test before assembly
	Total	330	33 seconds per wire

Materials

1. Solder: H65S-W1.6
2. Snap terminal: Based on sample
3. Lead wires: Based on sample
4. Flux: 70°C melting point (manufactured by *A* Co.)

See samples of all materials.

Special notes

1. The inspection consists of a tensile test on the soldered snap terminal (before cooling and before the plastic terminal has been screwed).
2. Be sure to describe any abnormal parts or operations and suggest improvement points.

Figure 5-18. Example of Standard Operation Sheet (for Soldering)

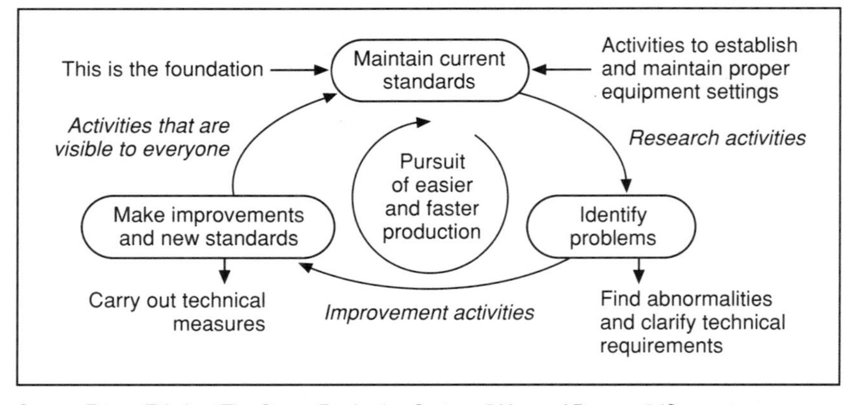

Source: Tatsuo Takaku, "The Canon Production System: 5 Years of Progress" (Corporate Organization Improvement Series), *IE Magazine* (JMA), November 1981.

Figure 5-19. The Source Control Wheel: An Unlimited Method for Achieving Zero Defects

Activities aimed at maintaining standards are more effective when their focus includes the following points:

1. *Are the standard sheets posted where they can be read easily?* Operators should be able to refer to them easily and managers should be able to check them to see how their measured results compare with the standards.
2. *At the very least, operations should be checked once a day against the standards.* If the standard operations are too difficult or if defects occur even when standards are maintained, a manager should be contacted immediately and corrective measures taken if necessary. Any improvements should be noted in the margin at the bottom of the standard sheet as important information. If the improvement is discussed but not written down, it may become part of the operators' know-how, but it will probably not become part of the company's assets.
3. *Improvement plans should be formally presented as improvement proposals.* Improvement plans should be discussed (and hopefully refined) by a small group, then drafted as formal improvement proposals. The more people involved in this process, the better; the improvement plan then becomes the employees' personal achievement and motivation to maintain any new standards that result in increases.

Step 7: Working toward Factory Automation

The basic rule of SN-type standardization is to work toward higher levels. Increasing automation of the factory represents a higher level of standardization that can help ensure the high-quality, safe, inexpensive, and timely manufacturing of the product. Automation should not put operators out of work but rather raise their skill levels and their abilities to contribute creatively to the company's success.

Figure 5-18 shows how standardization in soldering operations includes the development of a specialized soldering iron. Figures 5-16 and 5-17 show the analysis of many ideas for soldering improvements and the selection of a top few for implementation. With the improved soldering iron, the snap terminal connection site is preheated; solder is continually kept on the tip of the soldering iron that touches the lead wire and the parent material so that the thread-type solder can flow easily to make the connection between the wire and the snap terminal.

Once the snap terminal is positioned in its jig, the rest of the soldering can be done semiautomatically. After simplifying a process to make it easier, more reliable, and faster, changing the equipment to reduce labor requirements is relatively easy. Often such labor-saving improvements can be made inexpensively, since the new equipment is merely an aid rather than a substitute for human labor. Once an operator has organized an improvement idea on paper, other operators and improvement team members can help to develop the details of the plan. They may want to look into the advantages of buying commercial automated welding equipment, such as MIG or TIG machines.

In this pursuit, operators are going beyond their traditional roles and taking a new perspective as researchers. In today's era, more and more operators are cultivating this higher perspective rather than merely carrying out instructions. Studying work-site operations with their coworkers, operators are discovering better methods and are raising the quality of their work. Indeed, factory-floor workers are even leading the development of methods to build higher levels of reliability into products, an important theme in today's technological progress. Factory managers should recognize and support this new role for front-line employees as they themselves work toward standardization.

APPROACH TO ZERO DEFECTS

Having emphasized the importance of establishing and precisely maintaining standards, it is now time to emphasize that standards should be understood as nothing more than the foundation from which to build further improvements and a higher level of standards. No matter how much work is put into developing and maintaining good standards, they eventually must change in keeping with new technological advances and new environmental conditions.

Perfect standards do not exist at any company. Conditions always change, and standards must follow suit. The steps for standardization (summarized in Figure 5-9) do not change as quickly. Is there a way to accelerate the analysis of work-site problems and speed up the standardization process? Let us address this question in terms of quality control, looking at an example that illustrates this issue.

Less Analysis, More Improvement

Assembly process defects were a perennial problem for a certain manufacturer of automotive electrical parts. In response to this problem, the company organized an internal quality control conference to plan a quantitative analysis of the defect problem. The study group formed at this conference spent two months studying the defect data. They handed out check sheets to workshops so that employees could gather quantitative data over this two-month period. At another quality control conference held two months later, the study group presented the data in a carefully composed report that included multicolored Pareto diagrams and other visual aids. The quality control managers decided to implement a daily check to make sure the quantitative data were recorded correctly. These data are shown in Figure 5-20.

The report included the following points:

- Recently, there were negative trends in damaged products and operational characteristics of products, so everyone was asked to take more care. However:
  - There were many instances of loose screws. The number of loose screws seemed to decrease, but the same chronic defect problems still existed.

Process	Defective item	Month 1				Month 2				Total
		Week 1	Week 2	Week 3	Week 4	Week 1	Week 2	Week 3	Week 4	
Parts processing	Wire layout error		II	I	IIII	III		I		11
	Soldering defect		I	II		卌 III	卌 III	II		21
	Short circuit	I		I			I	II	IIII	9
Assembly	Loose screw	卌 II	卌 卌 卌 I	卌 卌 II	卌 II	卌 卌 卌	卌 II	卌 II	卌	76
	Missing or detached part			I	卌				卌	11
	Glue leakage	I		II	卌		I	II	IIII	15
Inspection	Invisible operational defects	II								2
	Dirty odometer	卌	卌 I	卌	卌 III	III	卌 I	卌 III	卌 II	48
	Damaged part	卌 I	卌 I	卌	II	II	I	I		23
	Total	22	31	29	31	31	24	23	25	216

Figure 5-20. Defect Data Compiled at Automotive Electrical Parts Manufacturer

- There were many defective gluing points during the fourth week of each month, which suggests that the end-of-the-month rush was making workers too tired and busy to take sufficient care when gluing parts.
- The occurrence of dirty odometers increased. The factory managers need to patrol the relevant processes more closely and more often.

The study group's approach could have been improved in the following ways:

1. Why did the study group spend two months simply gathering data? They should have taken a more direct, hands-on approach in checking out the problems immediately, using the 5W1H approach to ask, "What are the problems?" "Where and how often do they occur?" "Which equipment units and people are involved?" and so on. They should have

immediately identified the screws that came loose and which sections of which parts got damaged.

2. Because the problems occurred at the work site, the study group should have gone to each relevant process to check current conditions and find out why problems were occurring there. (In particular, they should have used an SN problem-finding checklist such as the one shown in Figure 5-10).
3. The study group should have tested a new set of countermeasures each week and each time a problem occurred. They should have also divided the problems into types such as easily correctable, large and difficult, and easily understood. They should have revised the standards after solving each problem and then checked up on the new conditions.

Because this company's improvement group took an inappropriate analytical approach, they spent two months wondering what to do while the factory turned out 216 defective products. If they had taken the SN approach, they would have responded to each problem with the improvement activity cycle of "discover problem → identify causes → make improvements → revise standards." That way, the factory would not have suffered the wasteful production of 216 defective items.

Presenting the data they had compiled at their office at a quality control conference was also a waste of time. After all, the purpose of the quality control conference was to reduce product defects, not to have more conferences. The only legitimate reason for holding a second conference two months later would have been to present reports on the results of improvements and to help other departments in the company make similar improvements. Conferences that only offer "postmortems" and "opinions" are useless.

Because problems occur at the work site, the key to gaining a quick understanding and finding a quick solution is to investigate the "scene of the crime," just like a police detective. To survive in today's competitive era of rapidly changing product models (some factories seldom make the same products two months in a row), factories must adopt wide-variety, small-lot production. This kind of production only runs smoothly with an accelerated cycle of discovering problems, finding causes, implementing countermeasures, and revising standards.

If you take an inappropriate approach to analyzing work-site problems, you will invariably end up with piles of mysteriously produced, defective goods. You can and must eliminate this kind of wasteful production.

The improvement group members at this company later learned the error of their ways and changed to a more practical approach. Using SN, they found solutions to problems and dramatically reduced the number of defective goods. They also decided to postpone another quality control conference scheduled two months later.

Use the QC Tools

The QC tools for eliminating defects are not as easy to use as a pair of scissors, nor do they show how to solve every defect problem. Nevertheless, they describe current conditions in an easy-to-understand manner. Figure 5-21 lists these tools, their applications, and their methods of use. The vertical columns in the chart list the QC steps for problem solving. Pay special attention to the descriptions of how these QC tools are used. Note also the many unmarked areas where no improvement efforts take place. It is important to recognize that the QC tools are not just nice graphics for final presentations; they should be applied to solve the problems initially.

Use Mistake-Proofing Methods

Mistake-proofing methods prevent backsliding on quality improvements and are therefore useful for enforcing standardization. During the oil crises of the 1970s, Dr. Shigeo Shingo introduced mistake-proofing principles (poka-yoke) into the Toyota production system. Other works treat the subject in more depth,* but in this book, it will suffice to present an overview of the implementation of mistake proofing, illustrated in Figure 5-22.

There are three types of mistake-proofing devices:

- Devices that prevent defects from occurring
- Devices that issue alarms when incorrect work procedures are detected
- Devices that verify that correct work procedures are being performed

* See Shigeo Shingo, *Zero Quality Control: Source Inspection and the Poka-Yoke System* (Cambridge, Mass.: Productivity Press, 1986); Nikkan Kogyo Shimbun, Ltd./Factory Magazine, eds., *Poka-Yoke: Improving Product Quality by Preventing Defects* (Cambridge, Mass.: Productivity Press, 1988). — Ed.

QC Tools	QC Steps									General use	Additional checkpoints for good ideas — alternative improvement methods
	Identify problems	Study current conditions (detailed problem identification)	Infer causes	Check to verify causes	Brainstorm improvement plans	Evaluate and select improvement plans	Test improvement plan	Evaluate and verify effects	Standardize improvements		
1. Graphs	○	◎	○	○				◎	○	For gaining a temporal picture of changes and for making comparisons	Set standards and targets, then study causes of any deviation that occurs and make corrective improvements.
2. Histograms	○	◎						○		For understanding variations in data	Use the 5M categories to find the causes of variation. Then take the same approach described above.
3. Pareto diagrams	◎	○		○				○		For understanding which problems are most important	Use a check sheet along with the Pareto diagram. Take countermeasures that address the most significant problems.

4. Cause-and-effect diagrams			◎		○					For analyzing and identifying the causes of problems and suggesting methods of controlling them	Use the 5M categories and repeatedly ask "Why?" to gather facts while filling out the diagram.
5. Check sheet		○						○	◎	Facilitates the gathering and organization of data	Same as for Pareto diagrams.
6. Data stratification	◎	◎		◎				○		For grouping data to clarify differences	Set standard values, then investigate differences and their causes.
7. Scatter diagrams		○		◎						For studying targeted factors	Repeatedly ask "Why?" to clarify relationships among factors.
8. Control charts	○	○						○	◎	For understanding abnormal data trends	Study data trends as a time series to identify preventable causal factors.

◎ = Very useful ○ = Useful

Figure 5-21. The QC Tools and Their Uses

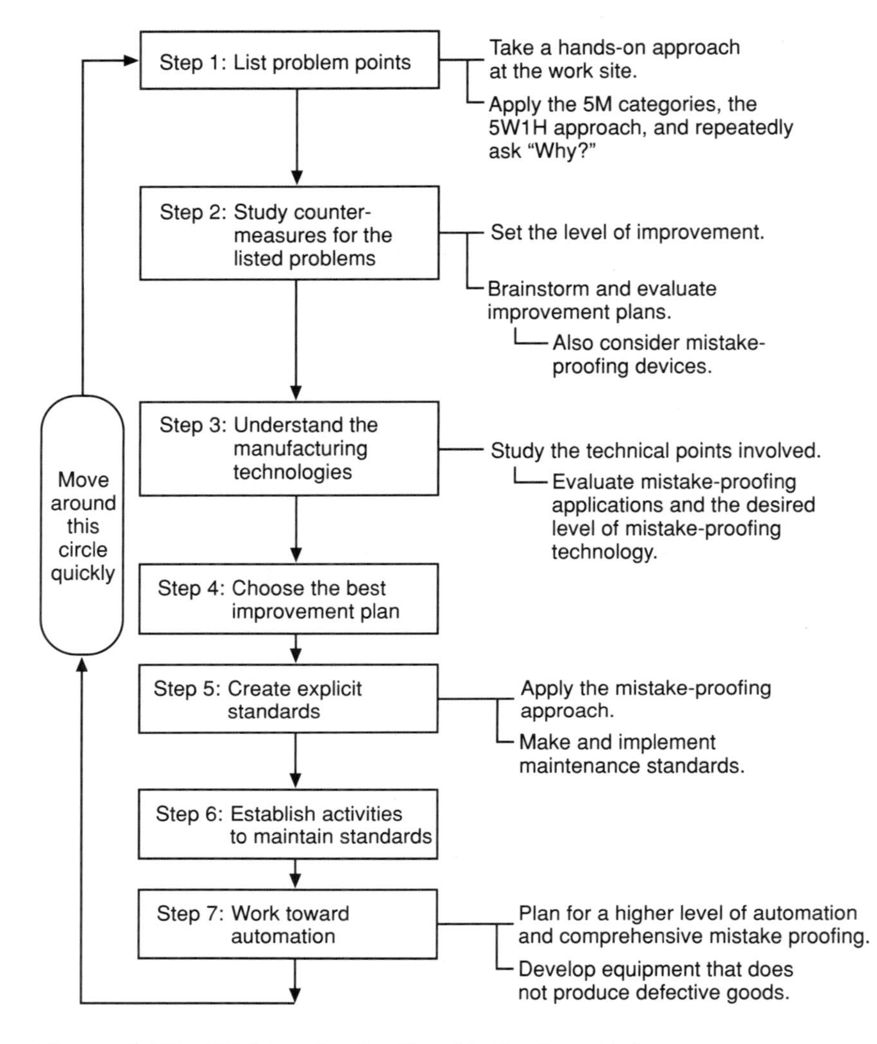

Figure 5-22. SN Standardization Methods and the Mistake-Proofing Approach

All three types generally use some sort of sensor, work reliably, and are inexpensive to make; often they are devised and installed by shop-floor workers. Good mistake-proofing devices that have proven themselves over long periods of use are invaluable aids to production.

6
Standardization for Equipment Management

WE HAVE AN INSPECTION CHECKLIST — WHY DON'T BREAKDOWNS DECREASE?

Most companies have established some sort of factory maintenance system, which often includes daily, weekly, and/or monthly inspections to help prevent equipment breakdowns. Equipment operators and maintenance technicians go through long checklists, which then go to the supervisor and section manager for approval, and eventually get filed away.

These procedures, however, do not prevent frequent breakdowns or at least frequent replacement of parts. Despite the obvious ineffectiveness of current factory maintenance systems, companies seldom reexamine the relation between equipment checking procedures and the occurrence of equipment breakdowns.

Some companies have responded appropriately by adopting total productive maintenance (TPM) or the 5S approach (industrial housekeeping activities). They consequently report cleaner equipment that breaks down less often and operates better for higher productivity.

One of the popular phrases that arose from improvement-minded equipment management is "cleaning is checking," another is "early detection is the key to preventing breakdowns." Companies would do well to reexamine their equipment maintenance systems from these perspectives. They should begin by asking:

1. How does the machine work?
2. What can be done to prevent equipment breakdowns?
3. What is the most effective way to use maintenance logs and checklists?

Finding answers to these questions is the first step in standardizing equipment management.

Taking Proper Care of Tools

Japanese swords and cutlery are famous for their high-quality craftsmanship. Indeed, their blend of strength, beauty, and usefulness make them one of the best products of Japanese civilization. The best Japanese swords and knives are made of a pure lustrous steel that is produced deep in Japan's mountainous San'in region using a foot-bellow foundry method.

Several famous master swordsmiths have operated these foot-bellows and forged blades from the special steel. The various processes involved in traditional Japanese sword making are displayed at the Yasugi Museum in the San'in region, along with photographs and descriptions of past and present master swordsmiths. Visitors sense the energetic spirit that went into the creation of those fine swords. The swordsmiths began and ended each workday with a prayer ritual for their tools and took meticulous care of them. The tools were stored gently in places where they were easily retrieved and used.

Although this tradition of reverent care for tools lives on as part of the modern Japanese work ethic, it has weakened over the years, and today's workers would do well to study the lessons of these great swordsmiths:

1. Each piece of equipment is a tool that produces things.
2. Our tools are our livelihood; we should therefore treat them with care and respect.
3. Every job includes three stages: preparation, operation, and clean up. Having reliable tools for the middle stage requires carrying out the first and third stages properly. The clean-up stage should include checking and maintenance to pave the way for the next job.

You can (and should) understand these points about production equipment. When equipment operators realize this, they will be less likely to think that equipment is too complicated for them to understand and maintain or to think, "I only operate it; someone else fixes it."

5S Activities Improve When Operators Understand the Principles Behind Breakdowns

Operator attitudes and actions regarding their equipment determine their responses to breakdowns. The following are three "musts" concerning the relation between operators and their equipment.

1. Operators must have a thorough understanding of their equipment.
2. Operators must understand the principles and causes related to equipment breakdowns so they can take corrective action.
3. Operators must learn and practice correct methods of operating and maintaining their equipment.

The behavior described above should be reinforced by reviewing a checklist of required actions daily. This greatly improves the way operators manage their equipment.

Figure 6-1 shows an example of such a checklist. One of the items on this checklist is "Oil supplied?" Seeing this item, an operator who understands the principles of and need for equipment maintenance would then determine precisely what to oil, when to oil it, why it needs oiling, how much oil to supply, and so on.

Referring back to the first "must" listed above, the operator should learn which parts of the equipment have fast and frequent vertical motions and therefore require high-speed lubricating oil and which parts move at slower speeds and lower temperatures and therefore require grease.

As for the second point, the operator should also know that if foreign matter, such as cutting chips, falls into the lubricant, it accelerates the deterioration of the bearings.

With regard to the third point, the operator should know how much oil loss is normal, how to check the oil level, and the possible reasons for obstructed oil supply in machines such as automobile engines (determined, for example, by using equipment that allows observation of oil flow during operation). The operator should also know how to take appropriate corrective action.

Since equipment and tools are made for people to use, it is only natural that the people who use them should understand their principles, main characteristics,

Job:	Early detection of breakdown potential in equipment
Category:	Rationalization of changeover Mistake-proofing Process linkage Layout Multi-process handling Reduction of processes Mechanization/Automation (Equipment maintenance)

Description of improvement:

Daily equipment inspections are carried out on steel can welding machines and painting equipment in the electronic products manufacturing department to reduce downtime resulting from equipment breakdowns.

Lathe A

1. Early detection of sections where breakdowns are likely to occur
2. Reduction of downtime due to equipment breakdowns

Machine tool checklist Equipment unit: Lathe A

Oper-ator → Group leader → Super. → Section mgr.

Group: Persons responsible for checking (Main:) (Alternate:)

Inspection symbols: No abnormalities; △ Caution point; ✕ Needs repair; ○ Repair completed

Equipment location: ______

Inspection Inverval	Check items	1	2	3	4	5	6	7	...	28	29	30	31
Daily	1. Abnormality in main power switch? (damage, melted part)												
	2. Abnormal noises in motor or pump?												
	3. Oil supplied?												
	4. Lubrication of surface friction areas?												
	5. Is oil within marked levels?												
	6. Overheating in main spindle or bearings?												
Weekly	1. Abnormal conditions in main power switch, fuse, or wire insulation?												
	2. Oil supplied?												
	3. Normal conditions in tool box?												
	4. Normal conditions for clamp?												
	5. Loose connections or buzzing in control panel?												
Monthly	1. ...												
	2. ...												

Figure 6-1. Example of Daily Equipment Checklist

and proper use. Many companies have recently begun intensive 5S (industrial housekeeping) activities for cleaning and organizing the workplace. Unless equipment operators follow the three "musts," however, their 5S activities are unlikely to be very effective.

When operators understand the principles behind equipment breakdowns (such as accelerated deterioration, described later), they first apply the 5S measures to the parts of the equipment that need them most. Operators who do not understand equipment principles aim 5S activities toward the most obvious areas (passageways, equipment covers, etc.); they often run out of time or interest before reaching the less conspicuous places that sorely need the 5Ss.

5S activities not backed by understanding of equipment principles will likely be ineffective as breakdown-prevention activities. Where an understanding of the equipment is particularly important but nonetheless absent, 5S activities are just another form of waste.

The three main activities for preventing equipment breakdowns are

1. Lubricating and inspecting
2. Cleaning (5S)
3. Preventing operation errors

You must determine why each of these activities is needed, where they are needed and to what extent, and who will carry them out and when. Once you understand these things, you can set your direction for standardization.

Figure 6-2 illustrates the bathtub curve, a statistical method that has long been used to predict equipment service life and the likelihood of breakdowns. With the advent of autonomous management and TPM, people have begun to question the validity of this statistical method. The bathtub curve may be valid in theory, but the activities described in the three boxes below the curve can greatly affect both breakdowns and equipment service life.

A certain factory conducted TPM activities, the results of which are shown in Figure 6-3. Although these activities took some time to complete, they nearly doubled the target line's productivity. Furthermore, very little was spent on new equipment; instead, a large number of small improvements were implemented.

This approach to improving equipment maintenance works in slow but steady increments, like an oyster making a pearl. When equipment has been mistreated and has deteriorated for ten years, it may take three to ten years to restore it to its original condition. Accordingly, it is much better to maintain equipment properly when it is new, thereby avoiding the long and laborious restoration process.

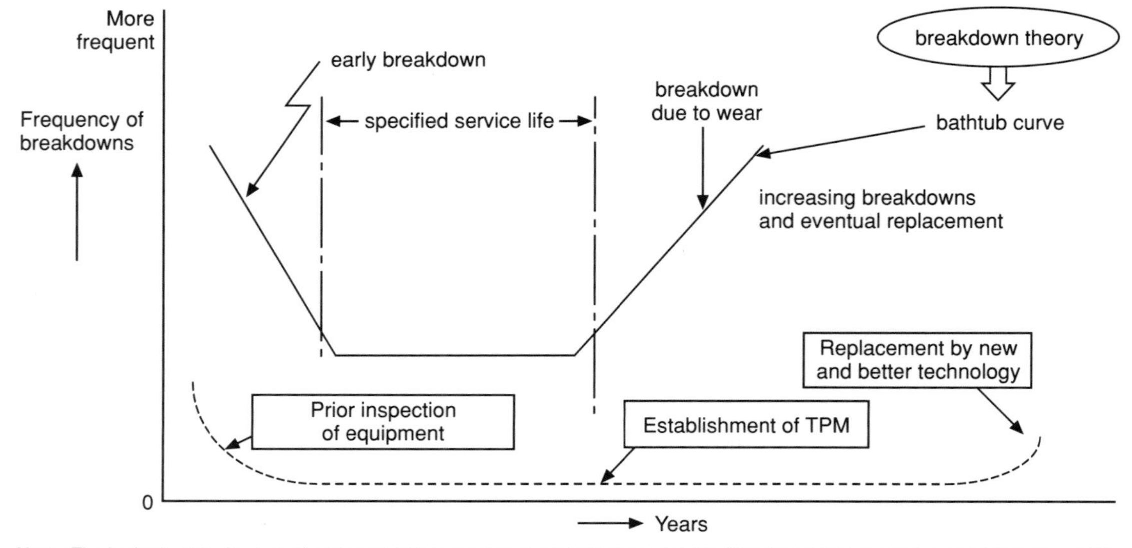

Note: The bathtub curve is an outdated approach to predicting equipment service life. The dotted line shows the curve that is possible using modern methods.

Figure 6-2. The Bathtub Curve

The Best Way to Use Equipment Inspection Checklists

When equipment has been restored to its full operating potential, you can then begin carrying out daily maintenance tasks. At this stage, it is essential for operators to take a sincere interest in keeping their equipment in top condition. Specifically, they must take special care in carrying out maintenance checks.

Equipment maintenance progresses through the following stages:

1. *Post-facto (breakdown) maintenance:* Equipment is expected to break down sooner or later. After it breaks down, we fix it.
2. *Preventive maintenance:* Equipment parts have a certain life expectancy, therefore we need to inspect and replace parts before their expiration date.
3. *Improvement (proactive) maintenance (small improvements based on abnormalities found during routine inspection):* Having detected abnormalities in equipment and equipment parts, we devise ways to prevent them from breaking down.
4. *Maintenance prevention:* We understand the life expectancy of parts and the proper use of equipment. We use methods to predict abnormalities, detect them early on, and prevent them.

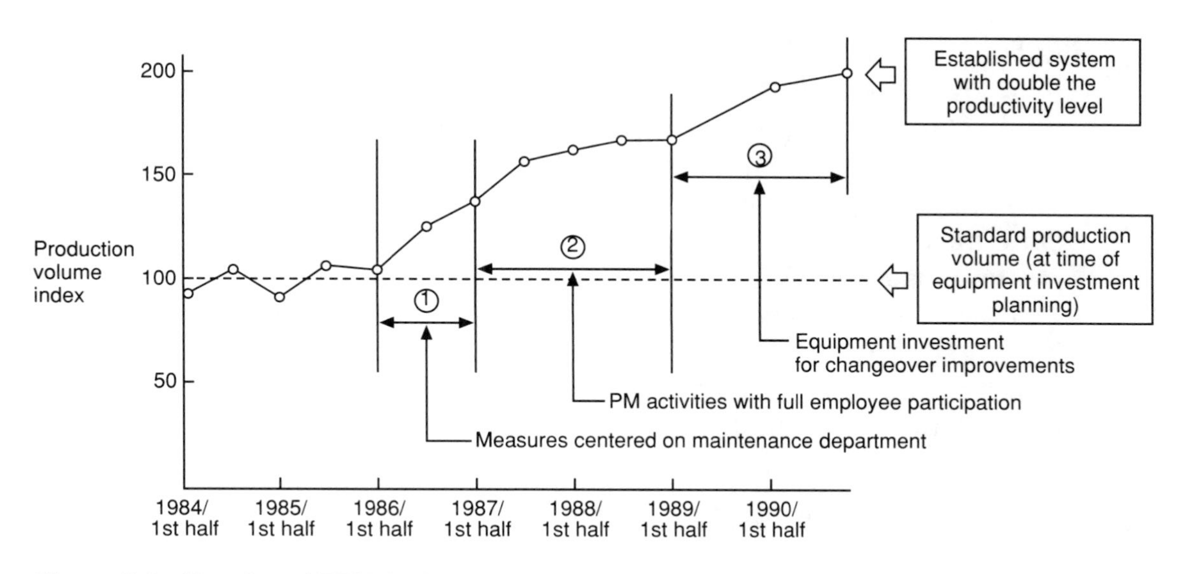

Figure 6-3. Results of TPM Activities

Proper maintenance checks are especially important in the fourth stage. Consider how the operators' jobs have changed under TPM:

1. These operators understand the structure of their equipment, the functions of its parts, and how these functions relate to one another.
2. As soon as the equipment begins producing defective goods or shows any kind of abnormality, they consider it a breakdown and immediately begin looking for problem points and devising countermeasures.
3. They carry out equipment maintenance checks as specified by the standards and stay in touch with maintenance technicians to ensure systematic maintenance. They make their own simple equipment repairs. They study maintenance skills they have not yet mastered and continually work toward a technical skill level they have established as a long-term goal.

Over time, shop-floor operators develop skills and activities that bring them closer to the role of a factory researcher. They begin to consider various aspects of the problems they encounter, gaining new perspectives on several TPM directives:

- *Cleaning is inspecting:* While you clean your equipment, you are also inspecting it. Conversely, it also means that when you fail to clean the equipment fully, you have not fully inspected it either.

- *Entering data in the maintenance log:* At stage four, operators do not take their log check items for granted. Instead, they ask "Why is this item necessary?" and "Is it possible to reduce the number of check items?" They also look for ways to make the checking procedures easier and less time-consuming, such as devising quick-release latches to simplify and accelerate the opening and closing of equipment covers during inspection, or creating inspection windows to avoid this motion altogether.
- *Turning in the maintenance log:* Maintenance logs can be used as memos to inform supervisors of abnormalities found during inspection. They can also be used to report on breakdown countermeasures and other improvements, or to gain feedback from superiors.

When this level is reached, inspections are effective for maintaining equipment in its optimum (full-capacity) condition. Routine equipment checks should entail more than running down checklists and filling out maintenance logs to detect omitted checkpoints. They should also involve the operator's five senses and personal pride in taking excellent care of the equipment. Ideally, equipment operators should feel concern and have aspirations of skill and knowledge similar to what good parents feel about raising their children.

STANDARDIZATION FOR BREAKDOWN ANALYSIS

The three main points in responding to equipment breakdowns are:

1. Learn the principles behind the breakdown.
2. Learn the facts behind the problem.
3. Establish a program and specific methods for reducing breakdowns.

The next sections examine these themes in depth.

Learn the Principles Behind Breakdowns

The principles behind breakdowns usually have to do with what could be called the "three evils:"

1. Inadequate cleaning
2. Inadequate lubrication
3. Erroneous operation

Figure 6-4 shows how these three evils lead to deterioration and eventual breakdowns. Deterioration is the main cause of most equipment breakdowns.

One of the points the noted Japanese management consultant Masakazu Nakaigawa (creator of the skills-management approach) emphasizes most is that if we can gain control of even the minor defects, we can keep defects close to zero and can accelerate the equipment's operation rate to 100 percent.* Focusing on equipment deterioration as the main principle for breakdown countermeasures, Nakaigawa divides deterioration into the categories shown in Figure 6-5 and stresses the importance of preventing accelerated deterioration.

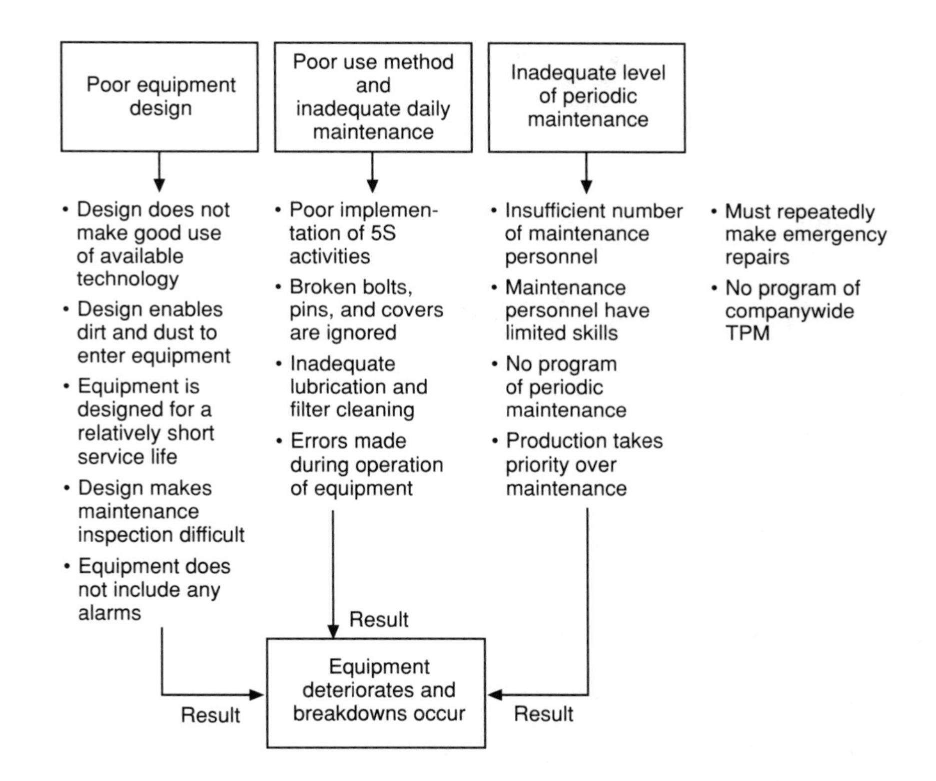

Figure 6-4. Weak Points That Lead to Equipment Deterioration

---

* Masakazu Nakaigawa, "Productivity Obstructions and Minor Defects," *Striving for Perfect Production via Video Skills Training* (Tokyo: Japan Management Association, 1957).

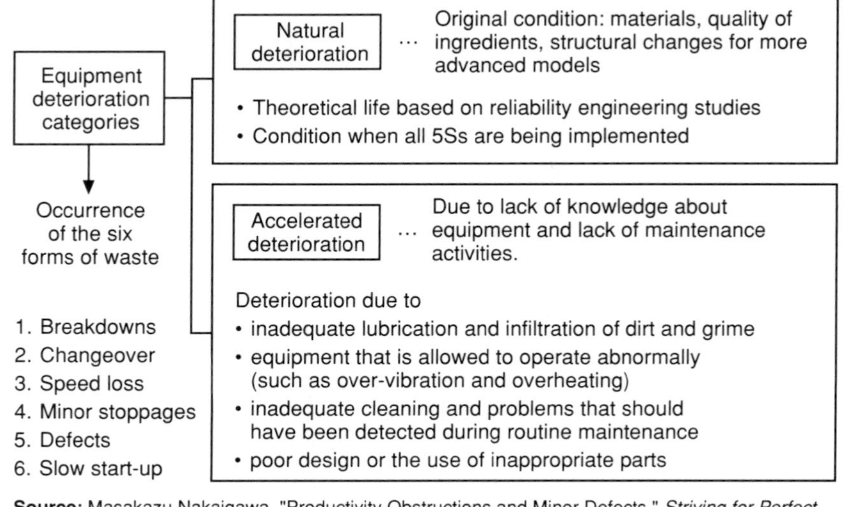

Source: Masakazu Nakaigawa, "Productivity Obstructions and Minor Defects," *Striving for Perfect Production via Video Skills Training*, Japan Management Association, 1957 (video).

Figure 6-5. Equipment Deterioration Categories

Many of the factories Nakaigawa has worked with have doubled their productivity with the same equipment and labor force simply by taking measures to control minor equipment defects. I have taken a similar approach in developing SN steps that have proved very effective for improving equipment capacity utilization. These steps are listed and described in Figure 6-6.

Addressing Breakdown Problems through Visual Management

When analyzing the breakdown of an individual piece of equipment, such as a time series, breakdown data records are a helpful standardization tool. Here we will study an example of how various facts related to frequent equipment breakdowns can be used in a visual management approach to analyzing breakdown problems and making improvements. Everyone can use this simple method effectively.

The first step in standardization for improving equipment capacity utilization is to get everyone to use the same methods for studying problems and making improvements. At the company where this case took place, I was working with the process management department, developing a computer-based production management system. No matter how good a production schedule I produced, the computer-based system always suffered some kind of break-

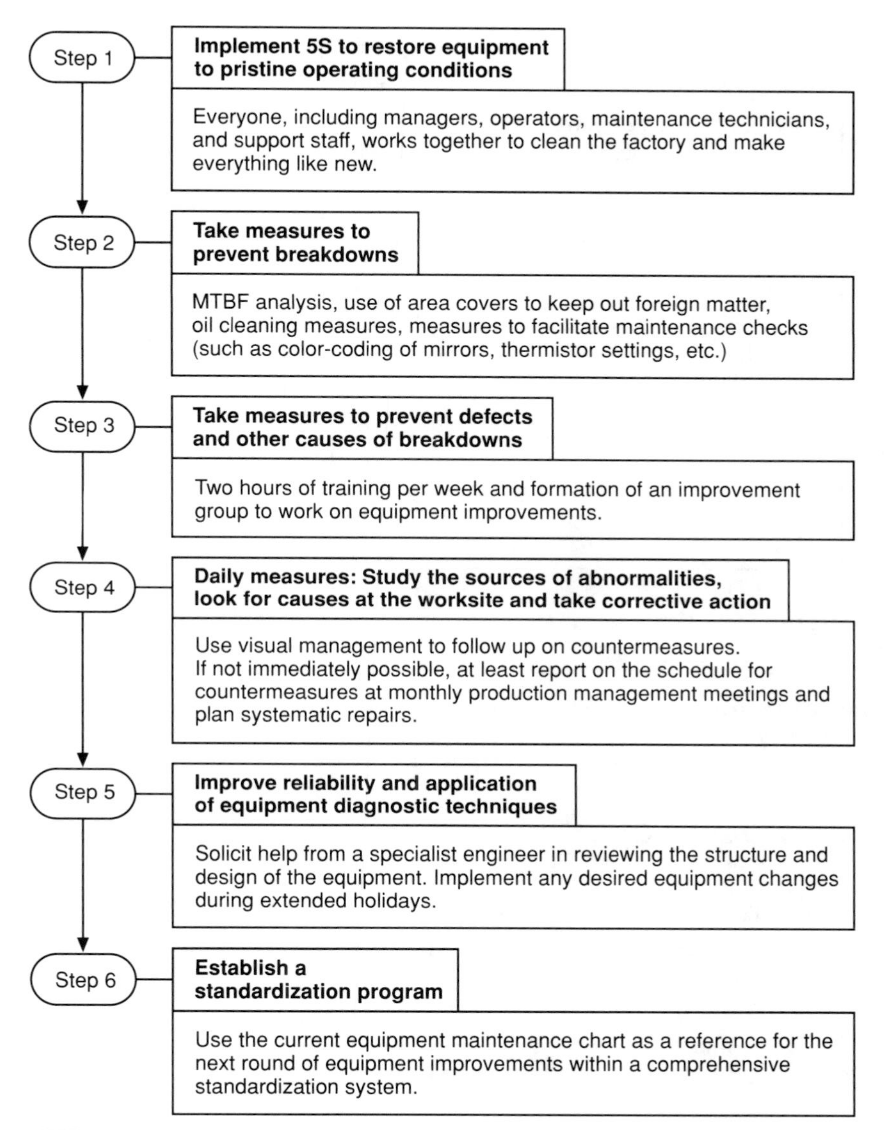

MTBF = Mean Time Between Failures (see also Figure 6-7)

Figure 6-6. SN Approach for Improving Equipment Capacity Utilization

down, causing the schedule to become disorganized. Instead of attempting to reschedule and risk another computer breakdown, I applied the MTBF principle from Step 2 in Figure 6-6 and took a visual management approach to the problem of equipment breakdown countermeasures (see Figure 6-7).

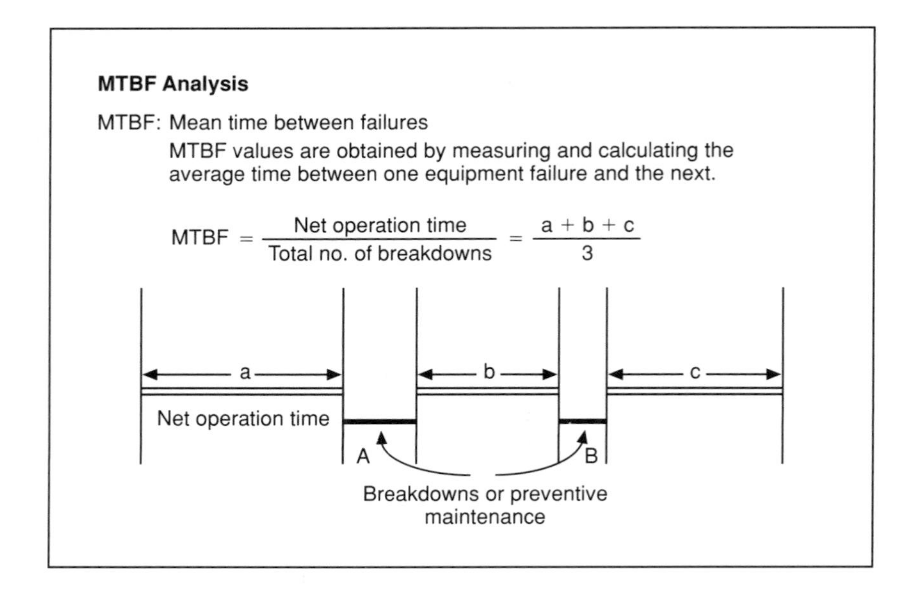

Figure 6-7. What Is MTBF?

Implementation of the visual management method involves the following steps:

Step 1: Create a Display Chart

Figure 6-8 shows an example of a display chart, in this case a visual management board. The equipment units are listed vertically and the time period (year and month) is noted horizontally.

Step 2: Make Color-Coded Maintenance Cards

- Red: Breakdown maintenance
- Yellow: Preventive maintenance
- Blue: Improvement maintenance

Star symbols indicate equipment that is currently down because of sporadic breakdowns or other reasons, such as changeover. A red star is used for sporadic breakdowns and a yellow star for other reasons.

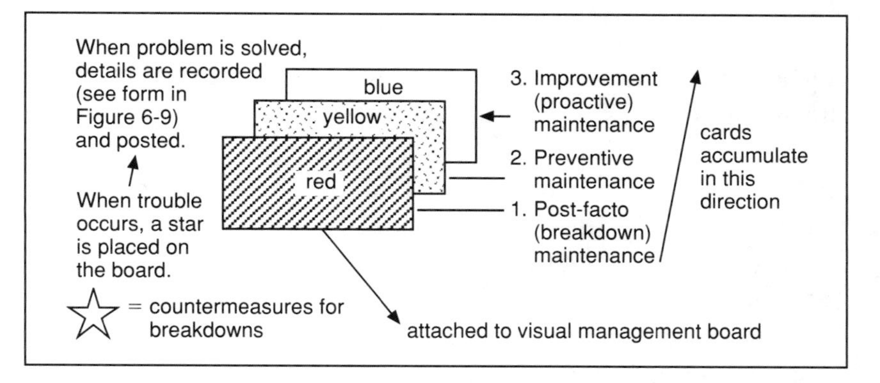

Example of a visual management board

card

Machine	April	May	June	July		Machine	April	May	June
A — 1		Breakdown				TD — 2	Improvement	Breakdown	Breakdown
AG — 1						HW — 1			
AG — 2						HW — 2		Improvement	
AG — 3						HW — 3	Improvement		
WH — 1	Preventive			Breakdown	☆	HW — 4	Preventive		
BT — 2	Preventive	Preventive	Breakdown			HW — 5		Preventive	Preventive
CP — 1	Improvement	Breakdown				HW — 6			
CP — 2						CW — 1, 2	Improvement	Breakdown	Improvement
SM	Breakdown	Preventive	Breakdown	Preventive		CW — 3			Improvement
CT			Preventive			B — 1	Breakdown	Breakdown	Breakdown
Ⓐ — 1						B — 2	Preventive		
Ⓐ — 2	Preventive	Preventive				CE — 1	Breakdown	Breakdown	Breakdown
Lift car	Preventive	Preventive				CE — 2		Improvement	Preventive
		Preventive	Preventive						

Breakdown maintenance | Preventive maintenance | Improvement maintenance | ☆ Breakdown or trouble

(The maintenance procedures carried out are described on each card.)

Figure 6-8. Visual Management Board (Kanban)

Step 3: Use the Display Chart and Cards

The cards display the required repair and maintenance work and the breakdown phenomena determined using the 5W1H approach. When a breakdown countermeasure has been completed, information about the problem and who took charge of it are transferred to the wall chart.

The cards are layered onto the kanban. If the equipment unit received breakdown maintenance, the red card must be placed on top, even if equipment improvement or preventive maintenance were also carried out.

Figure 6-9 shows the form used to request maintenance for "starred" (down) equipment and for giving work instructions. The figure also includes an explanation of how to use the form.

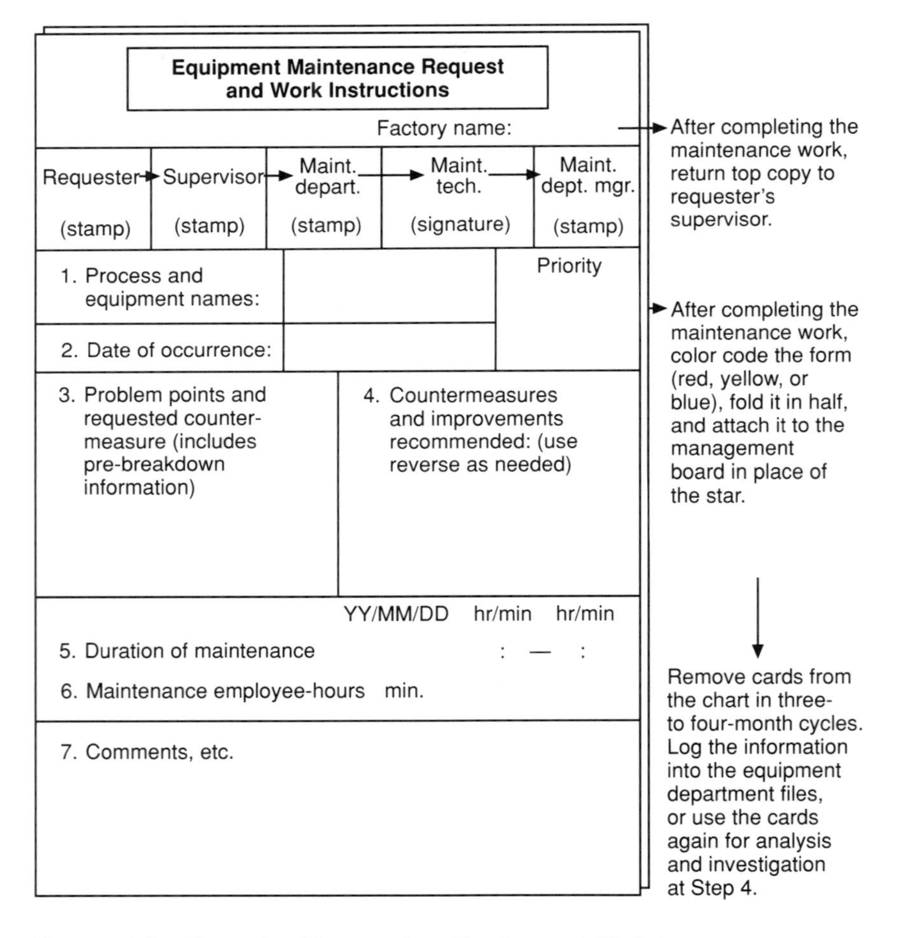

Equipment Maintenance Request and Work Instructions

Factory name:

Requester →	Supervisor →	Maint. depart. →	Maint. tech. →	Maint. dept. mgr.
(stamp)	(stamp)	(stamp)	(signature)	(stamp)

1. Process and equipment names:
2. Date of occurrence:

Priority

3. Problem points and requested countermeasure (includes pre-breakdown information)
4. Countermeasures and improvements recommended: (use reverse as needed)

YY/MM/DD hr/min hr/min

5. Duration of maintenance : — :
6. Maintenance employee-hours min.
7. Comments, etc.

Figure 6-9. Form for Requesting Equipment Maintenance and Entering Work Instructions

Step 4: Establish Priorities and Plans for Dealing with Equipment Problems (Establish Project Teams)

From a methodology perspective, this visual charting system uses the same type of methods found in the check sheets used for QC activities. A method similar to that described in Figure 5-21 is therefore useful here.

The project team should focus on three areas:

1. Equipment breakdowns that are infrequent but have a large impact on processes (deadlines) and meeting the production schedule.
2. Similar types of breakdowns that occur frequently.
3. Breakdowns that affect the smooth operation of other equipment in the process.

The project team is in charge of devising and prioritizing breakdown countermeasures. To help high-priority countermeasures go more smoothly, a second kind of visual chart is posted next to the management chart. This chart identifies each countermeasure as a high-priority improvement project and describes the planned action, its schedule, the people involved (project team members), and the required parts, tools, and materials.

Figure 6-10 shows an example of the kind of chart that describes project team activities. In response to frequent problems with bearings, the team carried out various analyses and found that too much grease had been applied to the bearings, resulting in grease leakage. When grease that has leaked from the bearings remains while the machine operates, metal and ceramic dust particles adhere to the grease. When grease is reapplied to the bearings, some of this dust gets into the bearings, resulting in accelerated deterioration.

The team's countermeasures include a thorough grease clean-up and inspection and replacement of grease caps. Thus, this countermeasure is targeted at operator behavior as well as equipment mechanisms.

Bearings fall into three classes: high-, medium-, and low-grade; each type was used where it was deemed appropriate. The team's research into bearing life and causes of defective products revealed that the process operations would generally proceed more reliably and quickly if high-grade, high-precision bearings are used in all cases. Therefore, they proposed using higher-quality bearings as part of their countermeasure plan.

For the friction-bearing parts and shafts, the team switched from their previous method of periodic lubrication to other methods, such as oil showers and oil baths. They found that even when using the same metal, the life of a part can vary drastically depending on whether the surface has been induction

Breakdown site	Target items and descriptive cards; analyzed every 2 to 3 months.	General counter-measure plan
Bearings	[cause] [cause] [ ] [ ] [ ] [ ] breakdowns/ big problems first ... other items grouped by machine type	Examples: Measures to reduce contamination; use of high-grade oil with the new high-grade bearings
Friction-bearing sections	[ ] [ ] [ ] [ ]	Examples: Improved lubrication method; use of heat-treated shaft material
Limit switches	[ ] [ ] [ ]	Example: Use of remote-control switches; reduced heat exposure

Figure 6-10. High-priority Countermeasure Chart with Color-coded Maintenance Cards

hardened. Therefore, they standardized the parts procurement vouchers by adding a check column for induction hardening.

As for the limit switches, the contact-point technology the team was using had a short service life, so they gradually switched to electrostatic capacitance technology. Limit switches have low heat resistance, and exposure to heat shortens their service life. Therefore, they devised a limit terminal that operates by remote control at a distance from the sensor.

These and other results from the team's experiments were displayed in a chart like the one shown in Figure 6-10. All of these improvements, which came from their investigation of problems on the factory floor, effectively reduced breakdowns.

Step 5: Follow Up on Effects

Using numerical values in tracking results of breakdown countermeasures is very helpful for two reasons. For one, it helps the project team members better appreciate the results of their efforts. For another, using numerical values to follow up on improvements enables you to judge whether your current improvement measures and policies are effective.

The data displays used for this kind of follow-up include:

1. Graphs showing equipment downtime
2. Graphs showing frequency of breakdowns
3. Other displays modeled on business performance indicators, such as loss figures.

Figure 6-11 shows a breakdown-frequency graph used during a follow-up. It was constructed from the cards on the display chart shown in Figure 6-8. This graph shows a large reduction from the initial level of sporadic breakdowns.

In this example, a total of 12 people were responsible for maintenance, conducting the work in two-person crews. Assuming 20 working days per month, we can say that one month contains 120 crew-days (20 days × 6 crews). Before improvement, post-facto maintenance jobs averaged 40 per month and accounted for 80 crew-days, while maintenance for improvement and routine maintenance checks required another 160 crew-days. As a result, maintenance crews worked overtime and holidays, and the company had to call for outside maintenance help. They tried but failed to come up with effective countermeasures to prevent breakdowns.

At this point, the company began using the card system to describe current conditions and enlisted the help of operators for repair work. This strategy gave new strength to the maintenance department and led to a string of improvements. In the end, the company achieved a two-thirds reduction in the frequency of sporadic breakdowns.

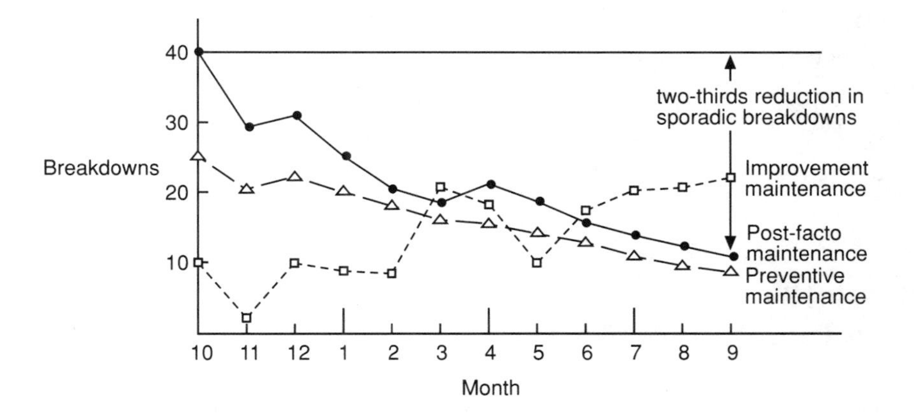

Figure 6-11. A Follow-up Chart Based on Color-coded Maintenance Cards

This experiment in using color-coded maintenance cards and wall charts to establish visual management involved a lot of skills practice and training, which bolstered employee awareness of equipment problems and ability to develop methods for solving them. The effects of this hands-on experience included the following:

1. Problems became clearly visible.
   - Everyone could easily recognize problems, and therefore institute countermeasures more quickly.
   - Everyone came to understand the overall balance of factory operations and how they relate to the priority of countermeasures and the degrees of difficulty.
2. Recurring problems were clearly identified.
   - All operators and supervisors asked why such problems kept recurring.
   - All members recognized the need to devise countermeasures for recurring problems.
   - Managers found it easier to issue instructions and to gain cooperation for their execution.
3. The team could confirm the fruits of the improvement efforts.
   - Having confirmed the effects of one countermeasure, they were better equipped to address the next improvement theme.

I once took part in labor-management negotiations as a union representative. From that perspective, I came to understand that although workers have unions that allow them to air grievances and work out solutions, no such union exists to spot machines in trouble and assist them. How can we hope to establish quick responses to equipment problems unless we first establish a system for making equipment problems visible and apparent? This realization led me to develop the display method previously described. The color-coded cards make it easy to see at a glance what kind of equipment problem exists and how severe it is.

ESTABLISHING A SYSTEM AND PROCEDURE FOR REDUCING BREAKDOWNS

To reduce breakdowns, the equipment must be inspected daily for early signs of common problems that cause failure. This requires establishing a system and procedure.

Figure 6-10 shows an example of a system of standard operations for carrying out equipment inspections. Inspection items should be few and simple, especially items checked daily. Simple check items are more accurate and tend to require more frequent inspection. Poka-yoke and alarm (*andon*) systems are useful for daily checks. It is also a good idea to arrange machine controls and layout to make equipment easier to inspect.

A similar approach can be used for more formal periodic inspections carried out weekly or monthly. Some factory managers take pride in conducting this level of inspection daily, but it is the quality of the inspection that matters, not the frequency. You should always try to eliminate maintenance operations that take too much time or are difficult or messy to carry out.

Figure 6-12 shows an excerpt of abnormality checkpoints from one factory's maintenance checklist. Figure 6-13 illustrates a factorywide maintenance system developed using an early problem detection approach.

Maintenance checks require good use of the senses as well as systematization. All employees should be trained to use their sensory perception not only to carry out safety improvement measures but also to predict where potential equipment safety hazards could occur.

Figure 6-14 shows an example of lubrication checkpoints. The checkpoints, numbered 1 through 7 in the drawing, are not described because the figure was originally used as a teaching exercise in which the students supply the descriptions. Equipment operators had 20 to 30 minutes to study the drawings, look for problems, and fill out the checkpoints. Afterward, the filled-out sheet was handed out and explained (see Figure 6-15).

This is an exercise in safety awareness training (SAT) for equipment. Chapter 3 discusses the use of an SAT approach with regard to risks to human workers. Here, similar principles are applied to the machines to safeguard their well-being and their continued effectiveness and availability. Requiring trainees to discuss and learn about the points raised in Figures 6-14 and 6-15 is a positive training method that is more effective than simply listening to an instructor's explanations. Operators who have been trained in this method return to their work sites with a more careful and concerned attitude about checking their equipment. Instructors should be encouraged to try this training approach.

A good equipment maintenance system applies the organizing principles shown in Figure 6-13 and a maintenance inspection approach, using a checklist like the one in Figure 6-15. A standardized system brings the equipment operators and maintenance technicians into a close working relationship, enabling

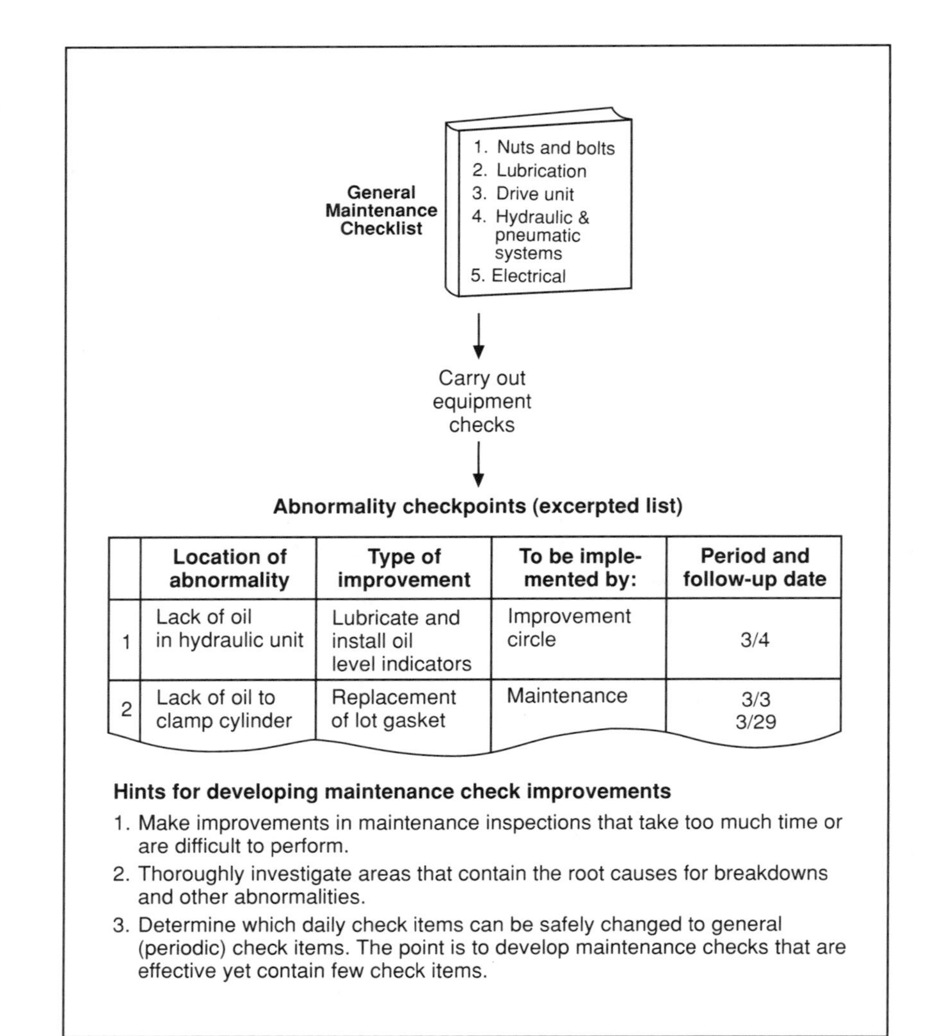

	Location of abnormality	Type of improvement	To be imple-mented by:	Period and follow-up date
1	Lack of oil in hydraulic unit	Lubricate and install oil level indicators	Improvement circle	3/4
2	Lack of oil to clamp cylinder	Replacement of lot gasket	Maintenance	3/3 3/29

Figure 6-12. Abnormality Checkpoints Excerpted from the General Maintenance Checklist

you to raise equipment capacity utilization to a higher level. This same approach can facilitate the smooth implementation of TPM-style companywide maintenance activities.

Rather than study only positive examples, let us also consider the following negative example and consider what measures can be taken to avoid such an outcome.

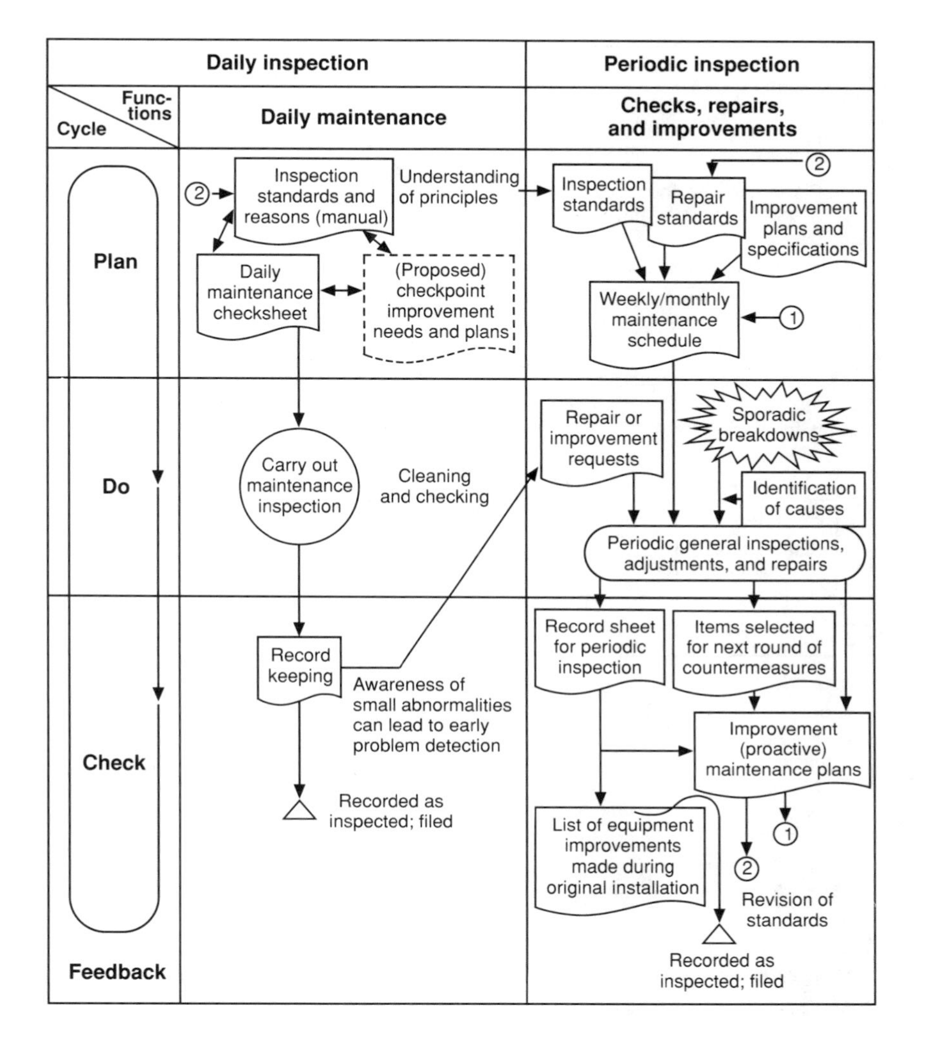

Figure 6-13. Equipment Checkpoints and Overall P-D-C Cycle

"We Have a Good Check Sheet, But We Still Have Breakdowns"

This situation happens now and then: Some checks were forgotten when the check sheet required a simple check-off for each item. One company switched to check sheets that required the entry of numerical values taken during the inspection. These check sheets were turned in and filed after each round of inspection.

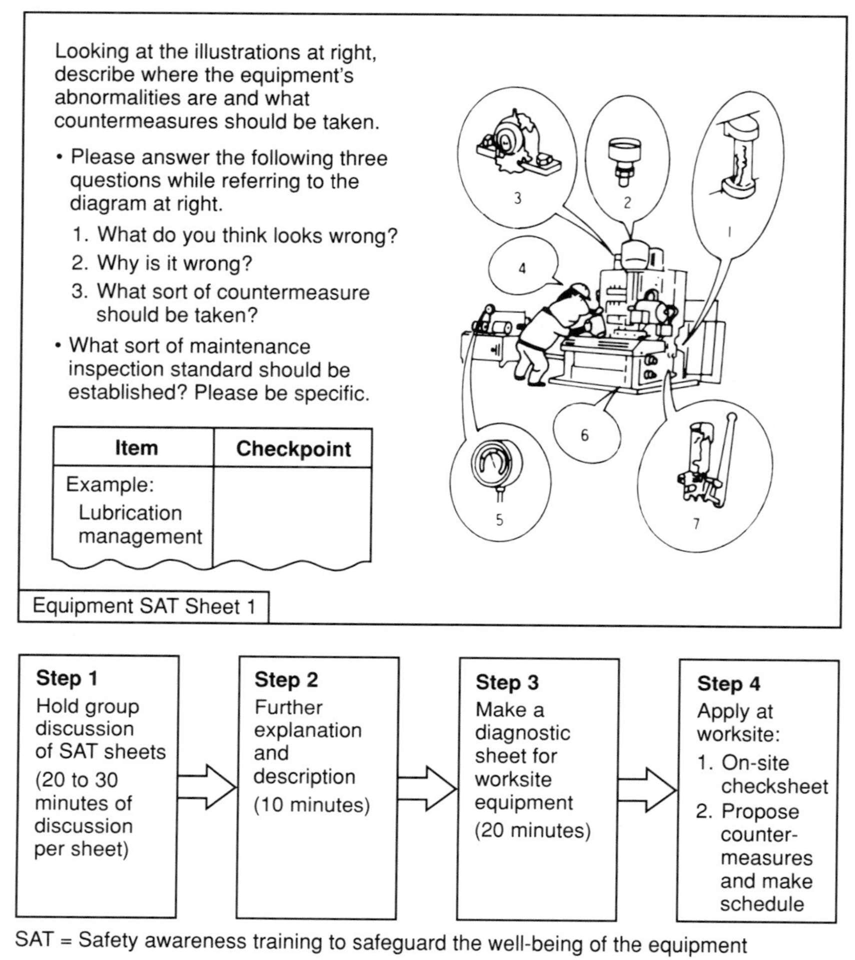

SAT = Safety awareness training to safeguard the well-being of the equipment

Source: *Our TPM* (Tokyo: Japan Institute of Plant Maintenance, 1982).

Figure 6-14. Equipment SAT Exercise Sheet

While investigating the cause of an oil pump breakdown, the workers found that the oil temperature remained above the standard temperature while the pump was operating. This caused the pump to overheat and burn, and the repair job required new parts. Unfortunately, this breakdown occurred when the factory was trying to get out an urgent order, and therefore it caused an uproar.

After the replacement parts came in, the company finally got things running again and shipped the urgent order, but the maintenance staff still wanted

Lubrication checkpoints

Lubrication is intended to prevent wearing and burning of friction-bearing parts and rotary parts and to prevent equipment deterioration. Every factory has many instances of defective lubrication that can lead to a shorter service life for equipment. Therefore, be extra careful when checking lubrication.

Item	Checkpoint
Lubrication management	1. Are the lubrication stands well-organized and clean? 2. Has an appropriate amount of lubricant been used? 3. Are the grease and oil cans well organized?
Lubrication unit	4. Can the gauge be checked easily to check the lubrication level? (Is it in a good position? Is it too dirty to read easily?) 5. Are tolerance limits indicated on the level gauge, and is the current level within limits? 6. Are tolerance limits indicated on the pressure gauge, and is the current pressure within limits? 7. Are there any strange sounds or excess heat coming from the pump or motor? 8. Is there any leakage of lubricant?
Lubrication equipment	9. Are auto-greasing or auto-oiling devices operating properly? 10. Are all branch valves operating normally, with oil circulating to each section? 11. Has the oiler become dirty on the inside or outside? Is the oil level correct? 12. Is there excessive grease or dust on the grease nipple? 13. Are there any defects on the grease nipple or cap? 14. Is there any grease leaking from the base unit? 15. Has excess oil spread to adjacent equipment? 16. Is there a proper layer of oil on friction-bearing parts and roller chains?

Source: *Our TPM* (Tokyo: Japan Institute of Plant Maintenance, 1982).

Figure 6-15. Answers to Lubrication Checkpoint Exercise

to know why the overheated oil had not been detected during a previous maintenance check.

Looking at the maintenance check sheets from the past few days, they found that oil temperature readings of 90 and 100° C had been entered. Obviously, these readings should have been recognized as indications of abnormality and something should have been done to prevent the pump burn-out.

What was the problem and what should have been done about it? You might start thinking right away about countermeasures such as mistake-proofing devices; however, first try to come up with an effective measure that does not cost any money. Consider the following three approaches:

1. *Identify a means of taking immediate action.* Assuming that the standard upper limit for the oil temperature is 95° C, you could have the maintenance check instructions say that a manager should be notified when a temperature reading of 95° C is obtained, and a maintenance technician should be contacted immediately for direct action if an oil temperature reading of 100° C is obtained.
2. *Flag the problem on the check sheet.* If the oil temperature is 95° C, attach a yellow sticky note to the check sheet. If it is 100° C, attach a red note to make sure the maintenance manager cannot overlook it.
3. *Hold a practice drill for daily maintenance and emergency procedures.* Operators should be drilled occasionally for maintenance emergencies just as they are for fires and other disasters. Alternatively, if the actions taken after detecting equipment abnormalities are written on the check sheet, have the operators practice taking those actions.

As long as you can make operational improvements to avoid the kind of problem described above, buying and installing mistake-proofing devices are unnecessary.

USING MAINTENANCE CHECK SHEETS FOR PLANNING NEW EQUIPMENT INSTALLATION AND SHUTDOWN MAINTENANCE

In a narrow sense, maintenance check sheets are for inspecting equipment conditions so that countermeasures can be taken for any abnormalities found. With this narrow application, however, you can only inspect and maintain the equipment at past performance levels. In any factory, people should also use their ingenuity to further improve the equipment, decreasing the check items on the list. This is *kaizen*, or continuous improvement.

One of the entries in the "check" section of Figure 6-13 is "List of equipment improvements made during original installation." The maintenance check sheet can be used to record this list; it can also be used for equipment improvements made during shutdown periods. In this case, the improvements are made under the "improvement [proactive] maintenance plans" also mentioned in Figure 6-13. By using maintenance check sheets to discover improvement points and plan countermeasures, it becomes clear how to use such check sheets for improving newly installed equipment and upgrading equipment during shutdown periods.

The general rule these days is to keep factory inventory to a minimum; the one exception to this rule is the factory's "inventory" of equipment improvement ideas — the more, the better. The best opportunities for increasing this inventory are during holidays or other long shutdown periods such as during factory layout conversions or new-equipment installations. Stay on the lookout for such opportunities and take full advantage of them when they arrive.

Let us examine two improvement-minded applications of maintenance check sheets: when installing new equipment and during long shutdowns.

New-Equipment Installation

As a practical matter, there are three reasons for keeping maintenance check records:

1. To estimate the service life of parts in planning periodic part replacements.
2. To acquire data for improvement-oriented maintenance.
3. To acquire data for supervision and other purposes.

The first two reasons are most important. When recording equipment maintenance data, try to avoid wasting time and energy entering data manually. The following methods are recommended:

- Use computer programs that index and process entered data to predict part service life and replacement periods.
- Organize past maintenance records to use when making improvements.

When planning equipment improvements, equipment designers rarely study past maintenance records. Usually, they just meet with the more experienced shop-floor people to get feedback on improvement needs. These meetings are usually formalities rather than true sources of improvement ideas. After all, the shop-floor managers deal with all kinds of problems, and it isn't easy for them to recall the equipment problems they noticed or ideas they had at one time or another. The equipment designer should at least prepare a checklist of possible equipment-improvement needs for the meeting. Going through this checklist item by item might help jog people's memories.

However, there are better ways to approach this problem, such as making sure the designer makes good use of existing drawings and specifications. These documents contain concise descriptions of improvement needs that equipment designers can easily understand and put to use in the next set of designs.

I recommend using monthly maintenance meetings and long shutdown periods to organize and update past maintenance records while discussing an improvement list with the maintenance staff. This method is outlined in Figure 6-16.

This method indicates clearly the technical history of each improved piece of equipment and can therefore help you avoid trying an interesting idea that had been applied before and was found unsuccessful. This method is not particularly easy to use; it requires teamwork among the shop-floor managers and the equipment designers. Nevertheless, no alternative is as effective.

Trouble should always be nipped in the bud, and making equipment as trouble-free as possible at the design stage is best. Therefore, it is worth reviewing data from periodic maintenance checks and selecting and arranging data related to improvement needs into charts for equipment designers to refer to the next time they visit the work site. The outline shown in Figure 6-16 is the basic pattern for this system.

Shutdown Maintenance

Most factories have a safety calendar showing the schedule of safety measures performed throughout the year. This schedule centers on activities undertaken by people; you should also consider making an "equipment safety calendar" focused on work-site maintenance for each unit.

When planning the equipment safety calendar and the parts replacement periods for the next year, you should refer to records of equipment improvements made during past shutdown periods. In fact, you should also summarize past periodic maintenance records and use those data as well for next year's schedule. This system for standardizing periodic maintenance and repair data can be very useful.

Standardizing the steps and items used for equipment repair and improvement work during long shutdown periods is also a good idea. Figure 6-17

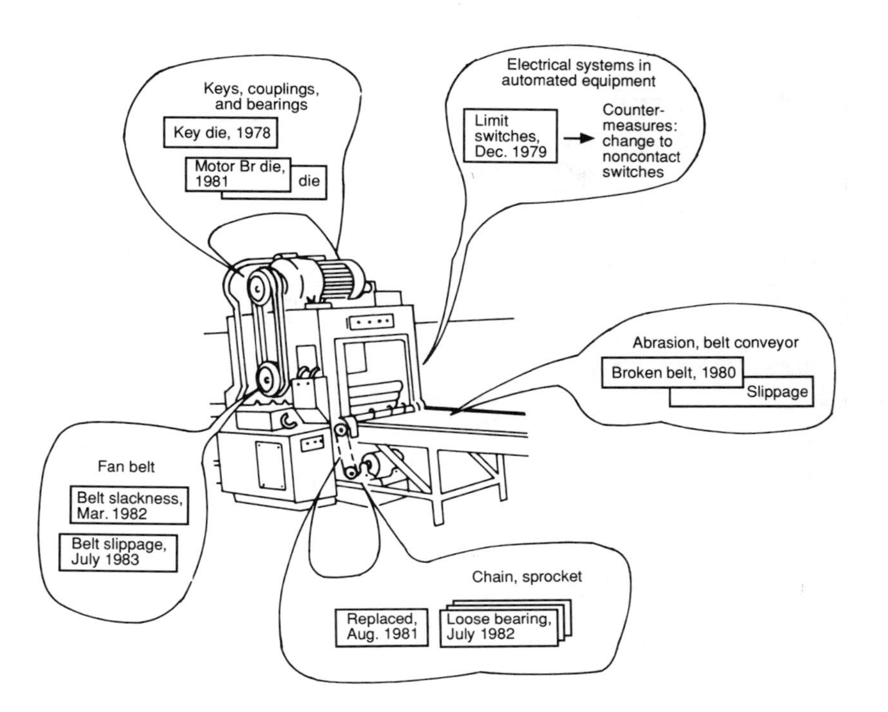

Notes about various malfunctions and breakdowns are recorded on a historical summary sheet for reference when new equipment is needed.

Figure 6-16. Equipment Maintenance Record Map

shows a standardized schedule of equipment repairs accomplished during shutdown periods. When the repairs are completed, use the same standards for the next round of equipment repair activities on the calendar, repeating the same steps each time.

This standardized method will help you avoid omissions as you prepare for the next round of shutdown repairs and will help the repair work go more smoothly. You should also consider applying PERT (program evaluation and review technique) when planning the detailed schedule of repairs, since this technique can effectively prevent errors and omissions in delivery of required parts and standardize the maintenance steps. PERT is also a good waste-prevention technique for planning maintenance steps and combinations of steps. Factories have used this technique widely when installing new equipment, changing the factory layout, or transferring equipment. Figure 6-18 shows an example of a maintenance schedule using PERT.

Period	Countermeasure items
1. Six months before shutdown	1. Appoint project team leaders. 2. Circulate memo among persons concerned to gather ideas. 3. Distribute list of improvement items.
2. Four months before shutdown	1. Start list of repairs desired during shutdown period. 2. Initiate contacts with outside staff. 3. Start list of employees involved in repair work.
3. Three months before shutdown	1. Meet with factory managers (group leaders) to divide list of desired repairs into the following categories: a. Repairs to be made before shutdown b. Repairs to be made during shutdown (broken down into repairs to be handled by in-house staff and repairs that require external help) c. Repairs to be made after the shutdown period. 2. Start contract negotiations with outside repair staff.
4. Two months before shutdown	1. Officially announce the repair categories and make charts. 2. Order and distribute required parts. 3. Confirm availability of items requested by outside repair staff (tools, welding equipment, rented rammers, etc.).
5. 1.8 months before shutdown	1. Select leader for each repair theme and make detailed schedule. 2. Start making detailed arrangements for each theme.
6. 1.5 months before shutdown	1. Review themes to look for possible combinations. 2. Reconfirm contents of repair themes to be handled by outside staff.
7. One week before shutdown	1. Confirm list of participants. 2. Meet with leaders to check schedule and make detailed arrangements concerning required materials. 3. Make start-up schedule for each process (detailed schedule for production start-up). 4. Display follow-up chart (including repair items, theme leaders, daily schedule, parts delivery schedule, list of participants, etc. for in-house and external staff).
8. Shutdown period	1. Safety first! 2. Review daily progress. 3. Keep records of problems and ideas found during the repair work (for example, repair work that took longer than expected, ways to lengthen part service life, structural improvements, ways to make maintenance and inspection easier, etc.).
9. Within two weeks after shutdown period	1. Review repair work in general and make lists and drawings containing reference data for next time. 2. Schedule remaining repair work and plan for follow-up. 3. List repair themes for the next shutdown period.

Figure 6-17. Example of Standardized Schedule of Equipment Repair during Shutdown Periods

Sheet No.:	Process/equipment: NC machine	Leader: Nakamura	Members: Yamada Nagata	Outside staff: Company A

Schedule						8/10	11	12	13	14	15	16	17			

Task

distribute parts
distribute tools and instruments
checklist of participants
final confir-mation
cleanup (5S)
shut off utilities
basic repair work
equipment disassembly
pipes
part replacements
equipment remod-eling
test
check electrical wiring
overall test
standby

Items to be considered beforehand

Problem	Improvement plan	Conclusion	
1. 8/10-8/12 Power shutoff throughout factory — no power equipment or tools can be used	Rent a small power generator, prepare parts before shutdown	Yamada	Ask supervisor to handle this — it's not a big problem.
2. Basic repair work and disassembly are done simultaneously. This presents a possible safety hazard due to falling parts or tools.	Install a safety net during cleanup.	Nagata	Can be used temporarily for all equipment except for type X.

Detailed implementation items

Item	Description	Method	In charge	Period	Check
1. Request for bearings	Order thrust bearings from Company A	2-month lead time — order soon	Nagata	7/15 Receive bearings 7/31 Check schedule 8/3 Check inventory	OK
2. Pipes	Two 4-meter sections of 1-inch pipe and 20 pipe joints	Prepare cutter and cut according to specs before shutdown	Nagata	7/31 Receive pipes	OK
3. Small generator	Rent from Company B	—	Nakamura	7/31	OK
15. Make start-up schedule	Display instructions on large wall chart and carry out test run	Same as left	Yamada	8/31	OK

Figure 6-18. PERT-Based Maintenance and Repair Schedule

STANDARDIZATION IS INDISPENSABLE FOR UNASSISTED OPERATION

Once you have turned equipment management know-how into standardized techniques, your next step is unassisted equipment operation. By "unassisted equipment operation" I do not mean fully automated equipment. Some tasks are still better left for people to perform. The greatest gains in productivity can be realized by developing an optimum relationship between operators and equipment.

People still outperform machines when it comes to the key characteristic of flexibility. Let us examine a case in point, where an operator was required to think flexibly about the job.

Must Someone Monitor for Minor Stoppages?

This case occurred in a highly automated parts-machining line. The line managers had planned on using a parts feeder to automate the supply of parts and materials to the line. At first glance, this device seemed to help the line operate more smoothly. The line workers had become skilled at multiprocess handling, and each operator was handling up to ten equipment units.

Minor stoppages began to occur occasionally on various part feeder machines. A stoppage would activate an alarm lamp and an operator would run over to fix the part feeders. Afterward, the line continued to operate automatically.

At one point, an operator developed the idea of running the equipment unassisted during the lunch hour. The plan was presented to the labor-management council, checked against the company regulations, and approved. The implementation of this idea began with a careful observation and analysis of current conditions. I was one of the factory managers at that work site.

The other managers and I watched the line operator at work and noticed how quickly and effectively he responded when a minor stoppage occurred. What would have happened, we wondered, if the stoppage had occurred with no operator around. In the meeting room, we considered allowing the machine to run idle when a minor stoppage occurred or installing a device that would automatically shut off the machine after a certain amount of time. Such ideas, however, were secondary to the main objective of dealing with problems such as minor stoppages in a way that enabled the equipment to operate without human assistance.

Personally, I thought that unassisted operation during the lunch break would be impossible unless we found some way to make responding to minor stoppages unnecessary. I came up with an idea that required the operator to change his thinking about the job.

1. The operator's approach had been to respond whenever a stoppage occurred. The operator's task had been recovery (repair).
2. To make unassisted operation possible required changing the operator's task to *identifying the causes* of minor stoppages. Instead of being a fix-it person, the operator must be a researcher, planner, and inspector of abnormalities.

This change was easy for the operator to understand in theory but not so simple to put into practice. For safety reasons, a barrier was erected in front of the machine, with a large sign warning that no one (except managers) was permitted to enter the area while the machine was operating unassisted. The managers shifted their lunch breaks so they could observe the unassisted operation while the operator was at lunch.

They also made a rule that, whenever a problem occurred, the person in charge would stop the machine immediately, check into the cause, and work with the equipment department in carrying out a countermeasure. After the first month of making one improvement after another, the machine ran completely unassisted for only about ten minutes a day.

During the second month, however, more effective improvements were made, so that the unassisted operation time increased to 1 to 2 hours per day. During the third month, the machine operator increased his understanding of the machine's operation to where he could present very specific improvement proposals to managers. Finally, things progressed to where the machine could be left to operate unassisted for as long as material supplies lasted. Within about 2.5 months, managers could stop supervising the machine during the lunch break.

This experience provided the following lessons:

1. *Automated operation required the operator to change his thinking*. The operator had to admit that the machine could not be left to operate unassisted during the lunch break as long as it required supervision. To overcome this problem, the operator had to change his attitude about his work and develop new skills as a researcher.

2. *Minor stoppages must be met with immediate investigation of causes and corrective improvements.* Such direct response and corrective action is much more effective and much less wasteful than holding meetings and discussing impractical theories.
3. *The managers must get to know the work site thoroughly.* In the process, they gain a better understanding of (and are better understood by) the work-site employees and the labor union. Managers and shop-floor workers can then work together more closely in developing automated processes.

Managers should always be aware that their assistance at the work-site must be a supportive, positive influence on the shop-floor workers. If managers fail to provide such an influence, they will have little success in their basic responsibility as managers to promote skills acquisition and improvement. Accordingly, managers should try to limit their assistance at the work site to one or two clearly defined periods during which they lend their wholehearted support.

Fortunately in the case described above, the work-site staff, managers, and labor union established a good relationship via frequent meetings, and as a result their joint project went smoothly. Such cooperation is essential for creating the confidence, technical skills, and well-coordinated improvement activities needed to develop unassisted operation.

Another important factor in developing unassisted operation is to stay alert for areas where unassisted operation seems possible and easy to accomplish. In a few cases the time and money required for achieving automation are inconsequential, but in most cases, automation schemes go nowhere unless they meet certain technical and financial conditions.

The table in Figure 6-19 was developed by Shigeo Shingo. It lists the steps toward automation, and as such can be useful when looking for promising areas for automation. It not only shows how human work is distinguished from machine work at each step, but serves as a clear set of signposts along the road to automation.

Chip Management Can Also Be Standardized

This topic may seem rather minor, but it deserves mention because cutting chips with automated equipment has become a very difficult problem. Fortunately, the problem of chip management can be reduced through standardization.

Type / Stage	Manual Functions				Mental Functions			
	Principal Operations				Marginal Allowances			
	Main Operations		Incidental Operations		(Usual Method)		(Toyota Method)	
	Cutting	Feeding	Installation/ Removal	Switch Operation	Detecting Abnormalities	Response to Abnormalities	Detecting Abnormalities	Response to Abnormalities
1. Manual operation	worker	worker	worker	worker	worker	worker	worker	worker
2. Manual feed, automatic cutting	machine	worker	worker	worker	worker	worker	worker	worker
3. Automatic feed, automatic cutting	machine		worker	worker	worker	worker	Machine that stops automatically (worker oversees more than one machine)	worker
4. Semiautomation	machine		machine	machine	worker	worker	Machine (worker oversees more than one machine)	worker
5. Preautomation (automation with a human touch)	machine		machine	machine	machine	machine	Machine (automation with a human touch)	worker
6. True automation	machine		machine		machine	machine	machine	machine

Step 1: Manual operations
Step 2: Manual feed with auto cutting (machine handles part of the work)
Step 3: Auto feed, auto cutting
Step 4: Semiautomation; auto setup and removal, auto feed, and auto cutting
Step 5: Automation of processing and detection of abnormalities. Machine is capable of automatically detecting abnormalities (preautomation).
Step 6: All processing, abnormality detection, and response to abnormalities are fully automated.

Source: Shigeo Shingo, *A Study of the Toyota Production System from an Industrial Engineering Viewpoint* (rev. ed.) (Cambridge, Mass.: Productivity Press, 1989), 70-71. (Originally published as *Study of "Toyota" Production System* [Tokyo: Japan Management Association, 1981].)

Figure 6-19. Human and Machine Work at Each Step Toward Automation

In discussing the automation of machining processes, we tend to focus on machining operations and the corresponding material handling and workpiece setup and removal tasks; however, no automated machining system is fully automated unless it also keeps the system clear of filings. The next sections, therefore, describe some of the chip management techniques I have obtained from my experience and show how they can fit into a standardization program.

Part of the problem with chips is that larger pieces can adhere to cutting tools and damage products. Generally, chip management issues fall into three categories: the shape of the filings, measures to remove the chips as they are produced, and removal of chip build-up in the equipment.

As for the first category, Figure 6-20 shows various types of chip shapes. Types E, F, and G are easiest to deal with, but soft materials usually produce chip types A or B. One very effective chip management technique is to install a chip breaker that breaks longer chips (A, B, or C) into smaller pieces (such as E, F, G, and H).

Figure 6-21 shows two examples of chip breaker designs for disposable cutting tools. Such chip breakers are important for joint research between cutting tool manufacturers and users. Researchers must standardize the dimen-

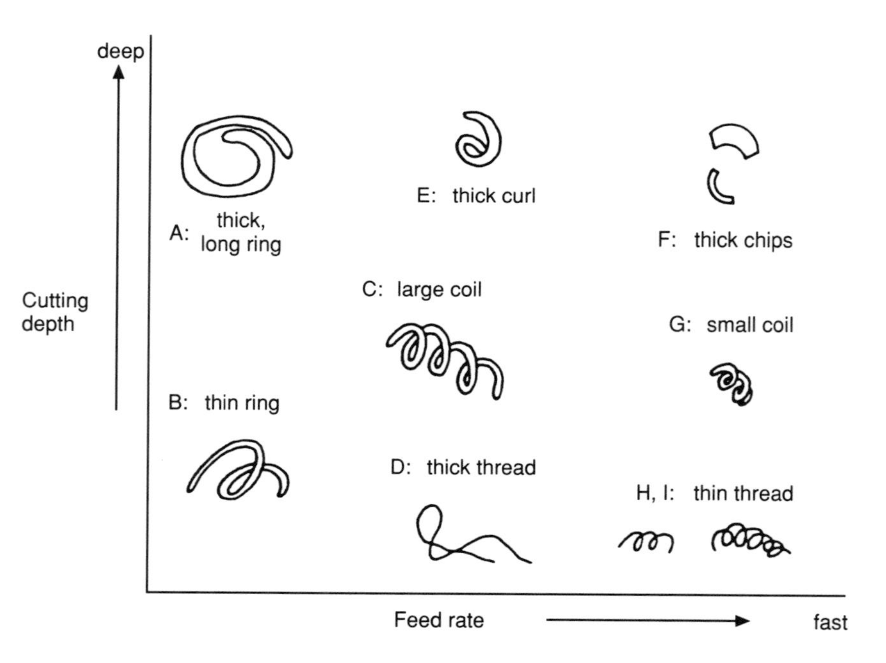

Figure 6-20. Chip Shapes Produced by Cutting or Drilling Metals

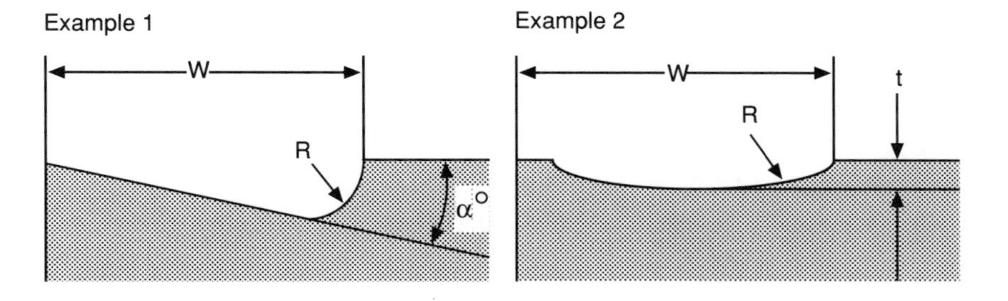

Figure 6-21. Chip Breaker Dimensions

sions shown as W, R, t, and α in the figure when designing the chip breaker shape according to the cutting conditions. After repeated cutting tests, they gain a detailed grasp of the relationship between the chip breaker shape and the feed rate, cutting depth, and chip shapes. Once these relationships are known, the chip breaker can be designed and standardized for the most productive conditions.

The second problem is that of chips scattering around and getting behind the cutting tool mechanism along with the cutting oil. When enough chips infiltrate cutting equipment this way, they can break or damage important parts, such as limit switches, timing belt covers, and wire covers. Effective countermeasures for this problem include establishing a smooth flow of chips mixed with cutting oil and using covers and curtains to prevent scattering. Rubber covers and curtains are often used, but they must be easy to replace since they tend to wear quickly. Door-sealing materials can be useful for this. We will return to the matter of quick and easy replacement later, when we discuss single changeover.

Removing chip types A, B, C, or D (see Figure 6-20) by conveyor is fairly difficult. Also, chips tend to accumulate in the grips of the conveyor and cause breakdowns. Some factories have responded to this problem by installing chip crushers for the conveyors, such as shown in part (a) of Figure 6-22. But when the cutting oil is returned to the cutting machine, small bits of chip still build up behind the conveyor. Therefore, if possible, it is better to have a pit or to elevate the cutting machine and use a chip breaker and simple removal system such as the one shown in part (b) of Figure 6-22.

You can map this experience using the model shown in Figure 6-16 and standardize it as the work-site technique used in the future.

Chip-related problems are indeed a headache, but once you organize them and use tables and graphs to plan suitable countermeasures, you are well on

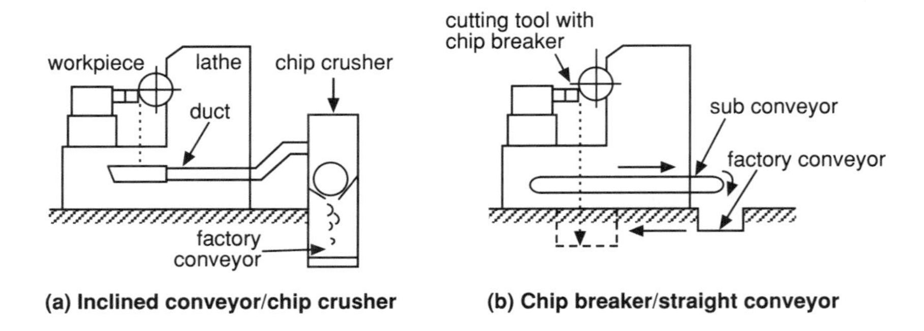

Figure 6-22. Improvement of Chip Removal

the road to overcoming them via standardization. Many readers will find it useful to learn more about different types of chip breakers. To learn more, consult handbooks that focus on this subject as well as materials available from cutting-tool manufacturers.

One problem related to management of cutting oil deserves mention here. Usually, cutting oil circulates through a filter system (such as the one shown in Figure 6-23) and is then reused. Such systems usually include a barrier wall and a filter. Although the barrier wall catches the larger chips, the smaller ones gradually accumulate on or near the filter. Eventually the filter becomes clogged, which leads to problems associated with a lack of cutting oil.

If you raise the standards in your approach to filters by thinking of filters as devices used to keep fluids clean, instead of devices that extract clean fluids from dirty fluids, you will pay more attention to how you maintain both the fil-

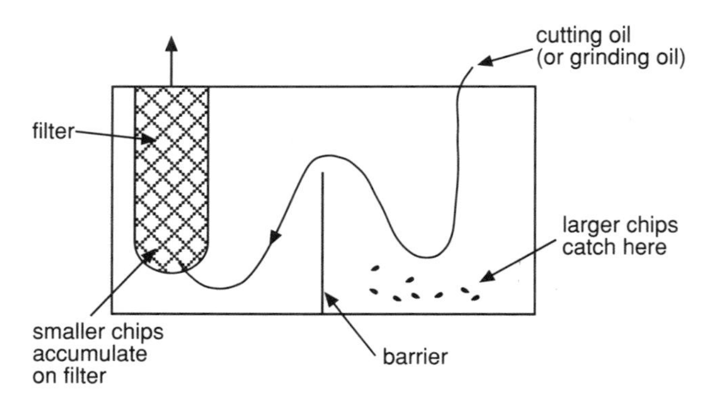

Figure 6-23. How Chips Accumulate in Cutting-Oil Filters

ter and the devices near it. You will carefully study the barrier, the depth of the filter tub, and the entire oil-cleaning cycle to find ways to prevent chips from accumulating on or near the filter. Once you find such methods, you need to standardize them, then see how they apply to other fluid-filtering systems, such as for lubricating oil.

MOVING TOWARD FACTORY AUTOMATION AND FLEXIBLE MANUFACTURING SYSTEMS

There are six major types of equipment-related loss or waste:

1. Breakdown loss
2. Setup and adjustment loss
3. Reduced speed loss
4. Idling/minor stoppage loss
5. Quality defect and rework loss
6. Start-up loss

Achieving zero loss in all of these equipment-related categories is an essential step in the development of FA (factory automation) and FMS (flexible manufacturing systems). Only when you have eliminated these losses can you begin to establish the conditions required for unassisted operation of equipment.

Figure 6-24 shows the various activities companies undertake as part of FA and/or FMS development programs. At first glance, quite a variety of techniques apparently are needed for such activities, in technical fields ranging from machining and assembly to computerized management systems. The following sections focus on two areas in which my experiences have provided some helpful techniques and caution points: single changeover techniques and equipment breakdown analysis.

Developing Efficient Single Changeover

Single changeover marks a radical departure from conventional equipment adjustment operations. In this context, the word *single* refers to a changeover completed in less than 10 minutes.

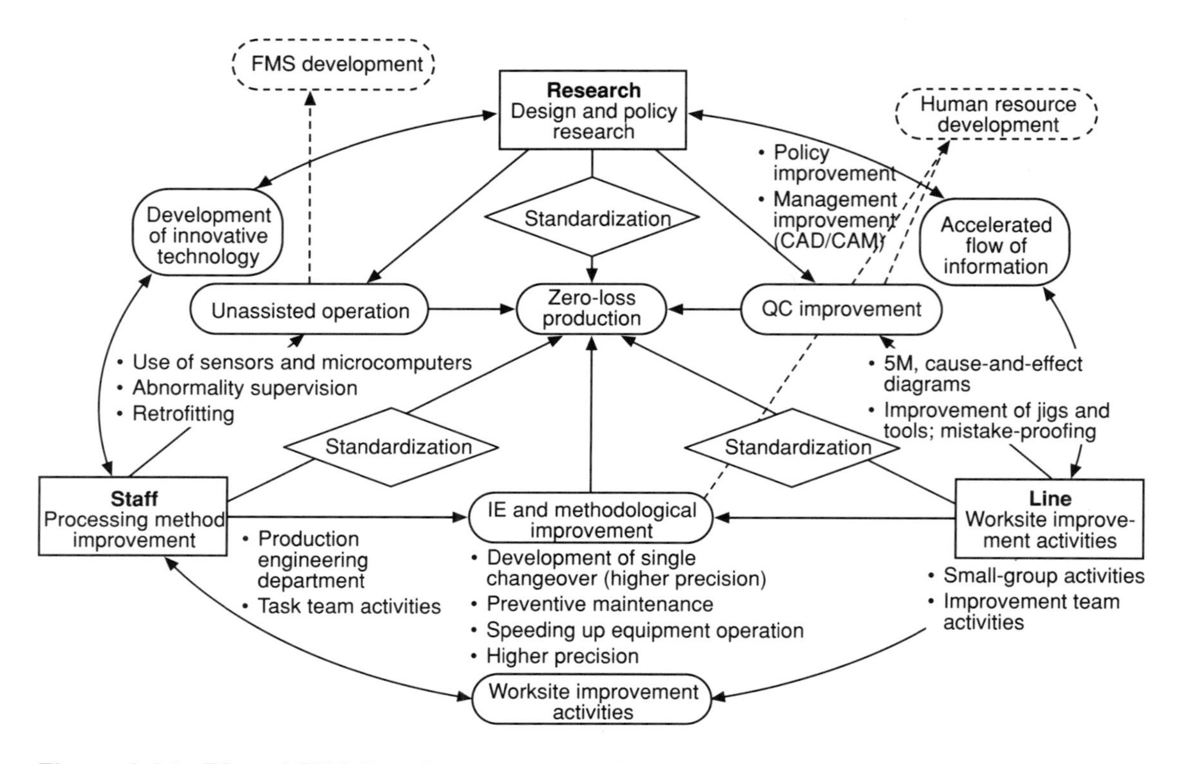

Figure 6-24. FA and FMS Development Activities

Readers who would like a full description of single changeover methods should read the writings of their developer, Shigeo Shingo.* Our discussion is restricted to several key points regarding single changeover, based on my experience in using these methods:

1. *Single changeover methods require some time-study research, but before doing such research it is best to set provisional standards.*

Time study usually requires many employee-hours, but you can speed up the process by making and using videotape recordings before beginning time study. I recommend the standard procedure in Figure 6-25 (pp. 154-55).

This approach to time study is less precise than it could be, but it does offer a standard procedure for carrying out operations. Continually making improvements during these operations brings you closer to single changeover. When you begin time study from this improved starting point, you spend less

* See Shigeo Shingo, *A Revolution in Manufacturing: The SMED System* (Cambridge, Mass.: Productivity Press, 1985).

time taking measurements and can keep identifying improvements areas even as you proceed with the video study and time study.

Using this video study method, one person makes the video recording while a second person takes time measurements. They list improvement needs and any improvement ideas that either have at the time. Later, they can replay the video recordings in slow motion to confirm the improvement needs and reevaluate their improvement plans. This is more efficient than doing the video study and time study first and then planning improvements later.

2. *When working out specific changeover methods, set clear targets expressed in functional terms before brainstorming for ideas.*

If you use functional terms (as in the VE approach) in describing specific work methods, it becomes much easier to identify the technical points requiring improvement; this will help you come up with improvement ideas. Consider, for example, the centering techniques illustrated in Figure 6-26, which can be described using the functional terms shown.

Ideas for improving positioning methods for die changeovers might come from such unlikely sources as bottle caps and coffee cup lids. Similarly, when looking for improvements in fastening down dies, which involve vertical, lateral, and inclined surface movements, you might consider ordinary consumer

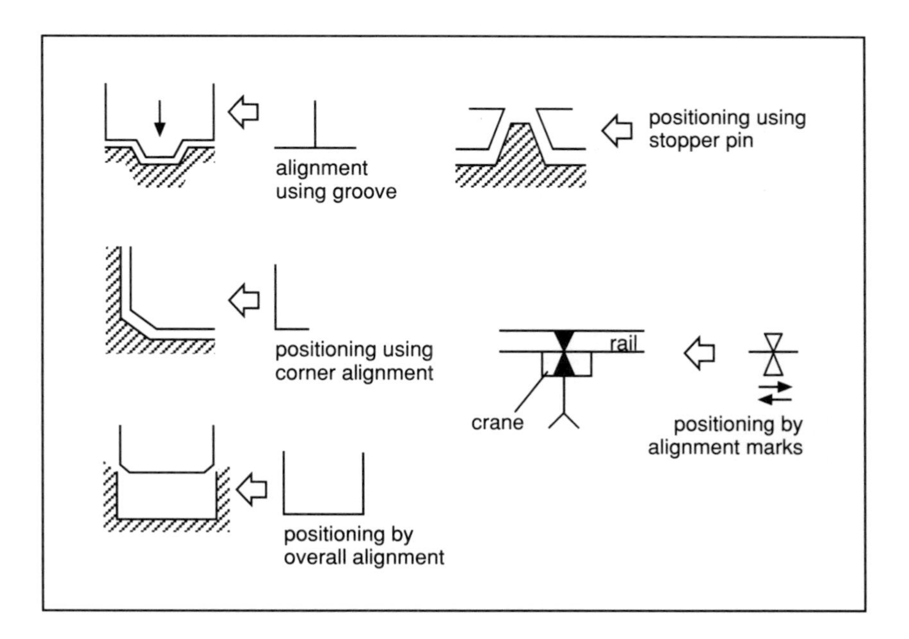

Figure 6-26. Centering Ideas Based on Function

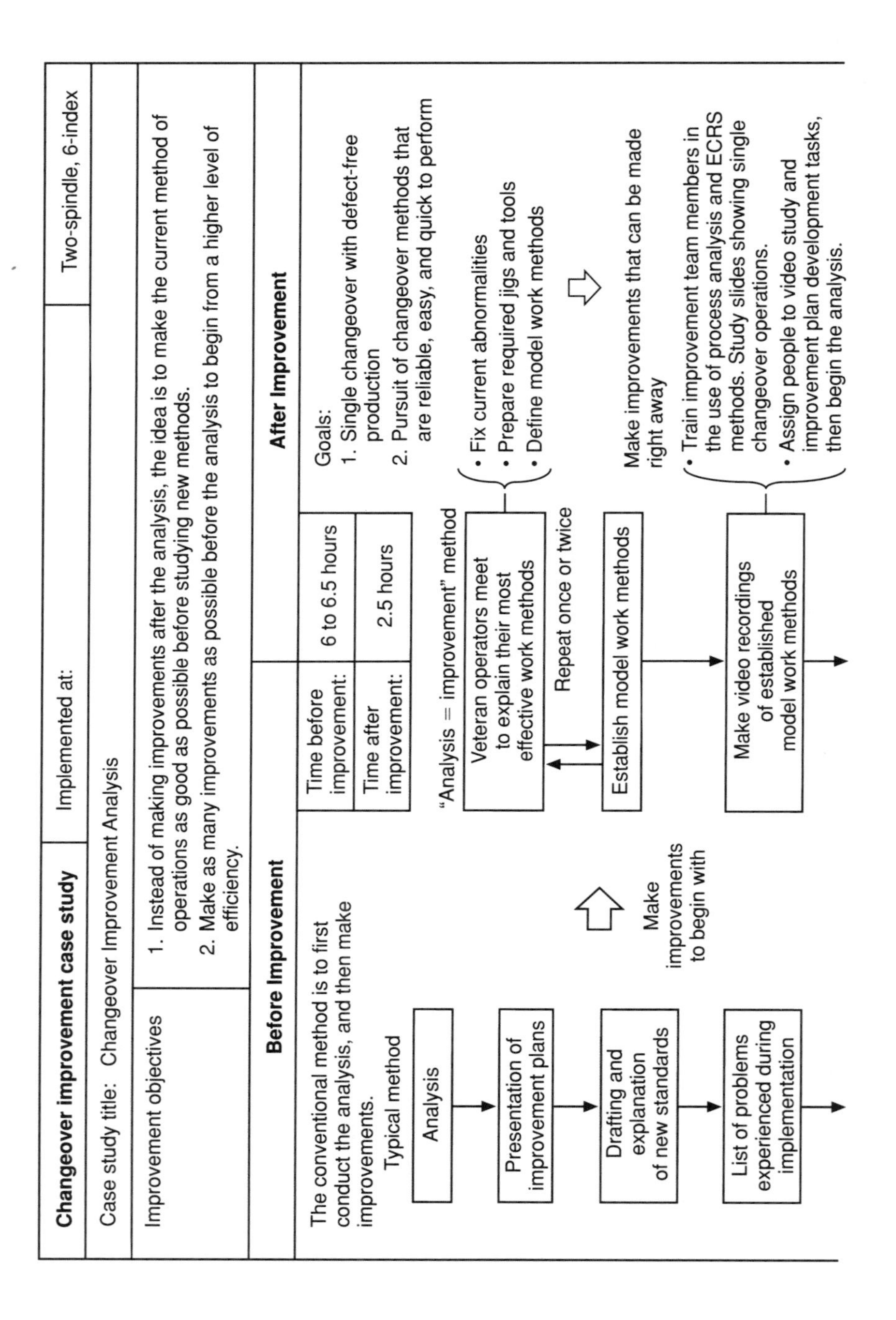
Changeover improvement case study
Implemented at:
Two-spindle, 6-index
Case study title: Changeover Improvement Analysis
Improvement objectives
1. Instead of making improvements after the analysis, the idea is to make the current method of operations as good as possible before studying new methods.
2. Make as many improvements as possible before the analysis to begin from a higher level of efficiency.
Before Improvement
After Improvement
The conventional method is to first conduct the analysis, and then make improvements.
Typical method
Analysis
Presentation of improvement plans
Drafting and explanation of new standards
List of problems experienced during implementation
Make improvements to begin with
Time before improvement:
6 to 6.5 hours
Time after improvement:
2.5 hours
Goals:
1. Single changeover with defect-free production
2. Pursuit of changeover methods that are reliable, easy, and quick to perform
"Analysis = improvement" method
Veteran operators meet to explain their most effective work methods
Repeat once or twice
• Fix current abnormalities
• Prepare required jigs and tools
• Define model work methods
Establish model work methods
Make improvements that can be made right away
Make video recordings of established model work methods
• Train improvement team members in the use of process analysis and ECRS methods. Study slides showing single changeover operations.
• Assign people to video study and improvement plan development tasks, then begin the analysis.

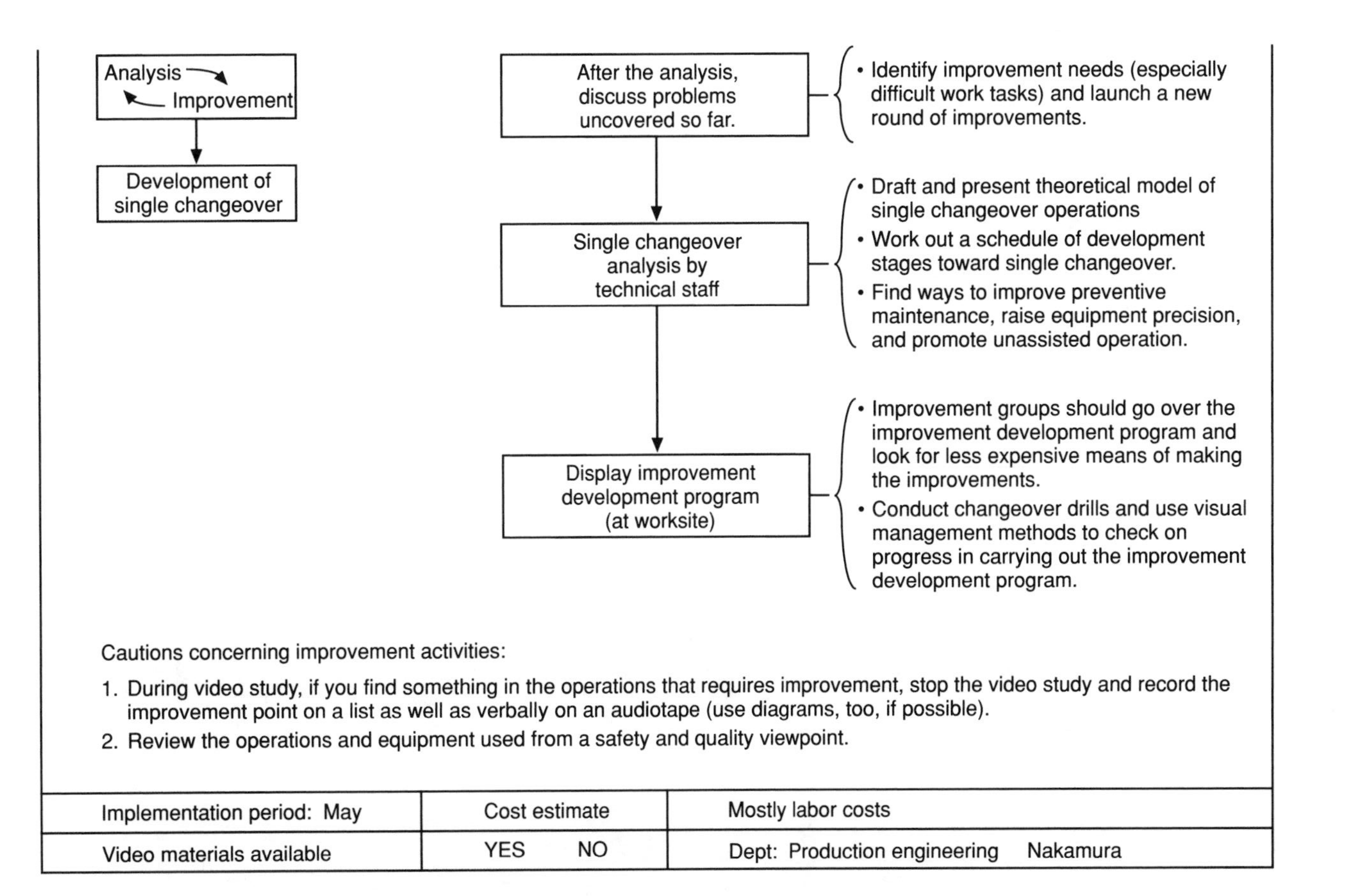

Figure 6-25. Video Study for Changeover Time Reductions

products, such as keys, sliding camera lens covers, and storage containers as possible sources of improvement ideas (see Figure 6-27).

3. *Study the mechanisms and structural principles other companies use in their single changeover development and borrow any ideas that seem appropriate to your own situation.*

Borrowing ideas from other companies includes not only taking factory inspection tours but also getting ideas from hardware, department, and even toy stores.

When you analyze these kinds of techniques using functional terms, you can see things you previously overlooked. You can also get a fresh perspective by discussing ideas with experts in different fields or by gathering information from product catalogs and other sources.

Breakdown Analysis Using Sensor Devices

Recent advances in computer technology have led to the development of various new kinds of sensors. Much research is being done in the use of these sensors for predicting breakdowns by detecting abnormalities in equipment. Already, sensor-based methods for predicting equipment breakdown have been standardized and application fields defined, as shown in Figure 6-28.

Suppose, for example, that you have a press that produces defective goods whenever its oil temperature rises, causing the pressing strength to drop, or a machine whose shaft has burned out due to poor circulation of lubricant. In either of these cases, temperature sensors could have detected

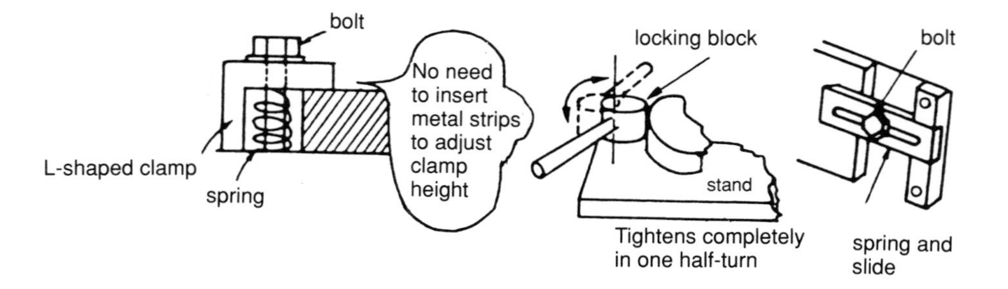

Figure 6-27. Die Fastening Ideas

such changes in oil temperature and provided warnings before defective goods were produced.

Figure 6-29 lists standards Nippon Steel's Kimitsu Works established for using equipment diagnostic sensors. Many Japanese steel companies now use these same standards.

Figure 6-30 describes several types of diagnostic instruments, which are often used to measure and analyze machine vibration. Figure 6-31 shows examples of applications of these devices.

The explanations and examples in this chapter may have given the mistaken impression that equipment standardization is useful only in the rather narrow field of routine equipment maintenance. This is certainly untrue, and I would prefer to leave the reader with an image of equipment standardization as shown in the figure below. In the figure, standardization and equipment improvement are the two legs on which quality and productivity improvement are carried forward. Meanwhile, one hand uses education and support to reach toward new technology, equipment, and products, while the other hand tests and verifies existing approaches. The eyes are focused on the targets as the entire body moves forward, step by step.

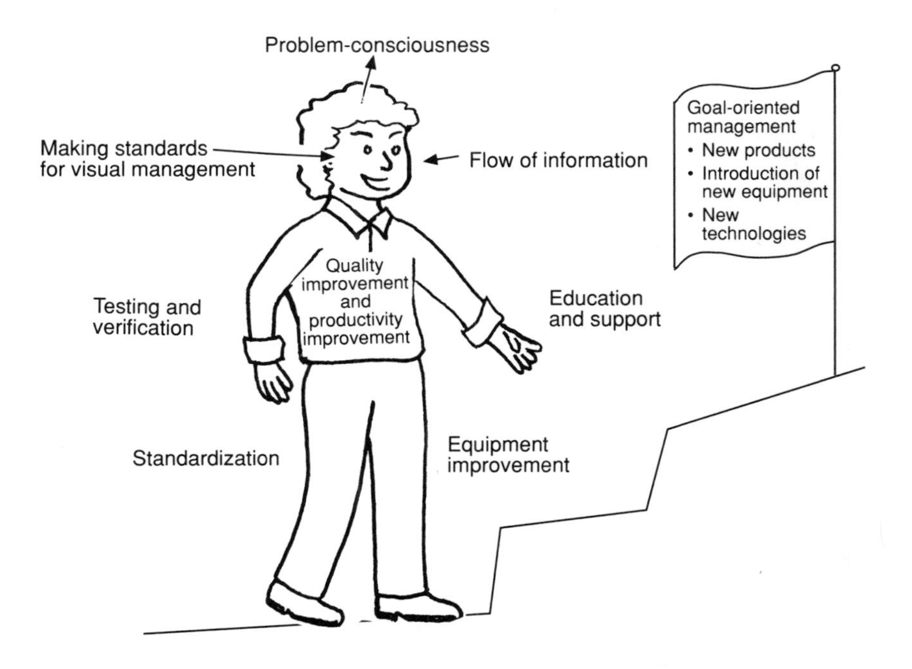

Detection method \ Detection function		Presence	Position checking	Dimensional		Flaws	Color differences
				General	Precision		
Contact	Limit switch	●	●	▲			
	Micro switch	●	●		▲		
	Touch switch	●	●		●		
	Differential transformer	●	●		●		
	Trimetron		●		●		
Non-contact	Proximity switch	●	●		●		
	Optical switch (linear)	●	●	▲			
	Optical switch (reflective)	●	●	▲		▲	●
	Fluid control device	●	●				

Detection method \ Detection function		Pressure	Temperature	Load (current)	Quantity	Time	Timing
Pressure gauge		●					
Pressure switch		●					
Meter relay				●			
Nugget tester				●			
Temperature gauges	Thermostat		●				
	Thermistor		●				
	Thermocoupler		●				
Counter					●		
Preset counter					●		
Timer						●	
Remote relay							●

Figure 6-28. Detection Methods for Breakdown Analysis and Prediction

Category	Device type	Function type	Examples	Main application
Checking	Checking devices	Multifunction	Machine checker	For checking conditions and monitoring trends
		Single function	Insulation checker, gear tooth checker, bearing checker	
Simple diagnostic tests	Monitoring devices	Component monitoring devices	Bearing monitor, shaft monitor	
		Function monitoring devices	Pop monitor, cylinder monitor	
		System monitoring devices	Lubrication system monitor, hydraulic system monitor	
		Plant monitoring devices	Blast furnace monitoring system	
High-precision diagnostic tests	Equipment monitoring devices	Multi-function	Rotary equipment diagnostic device, electrical equipment diagnostic device	Contact diagnostis, test-run diagnostics
		Single function	Insulation diagnostic device, spindle clutch diagnostic device, bearing diagnostic device, gear tooth diagnostic device	Contact diagnostics, test-run diagnostics, retrofitting, production analysis
	Diagnostic analyzers	Multi-function	Equipment signal preprocessing device, real-time analyzer	
		Single function	Transient memory device, average response device	

Figure 6-29. Various Diagnostic Devices for Equipment (Developed by Nippon Steel, Kimitsu Works)

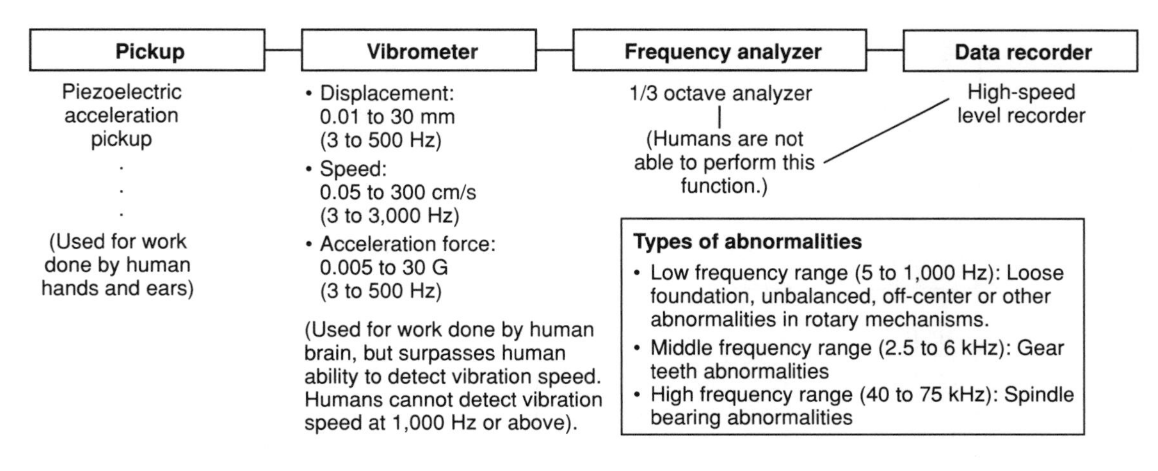

Figure 6-30. Characteristics of Equipment Diagnostic Instruments

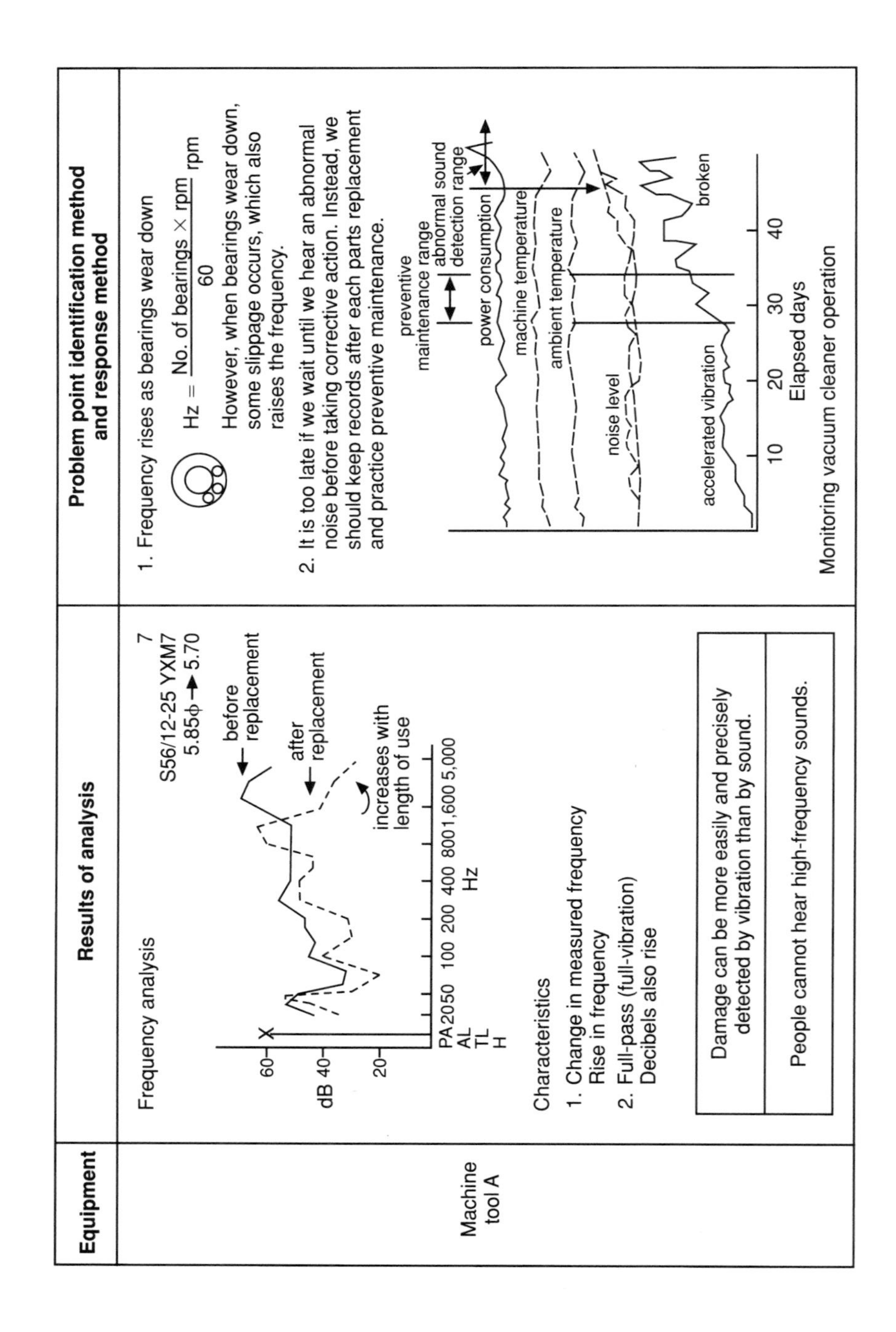

Equipment	Results of analysis	Problem point identification method and response method
Machine tool A	Frequency analysis Characteristics 1. Change in measured frequency Rise in frequency 2. Full-pass (full-vibration) Decibels also rise Damage can be more easily and precisely detected by vibration than by sound. People cannot hear high-frequency sounds.	1. Frequency rises as bearings wear down $Hz = \frac{\text{No. of bearings} \times \text{rpm}}{60}$ rpm However, when bearings wear down, some slippage occurs, which also raises the frequency. 2. It is too late if we wait until we hear an abnormal noise before taking corrective action. Instead, we should keep records after each parts replacement and practice preventive maintenance. Monitoring vacuum cleaner operation

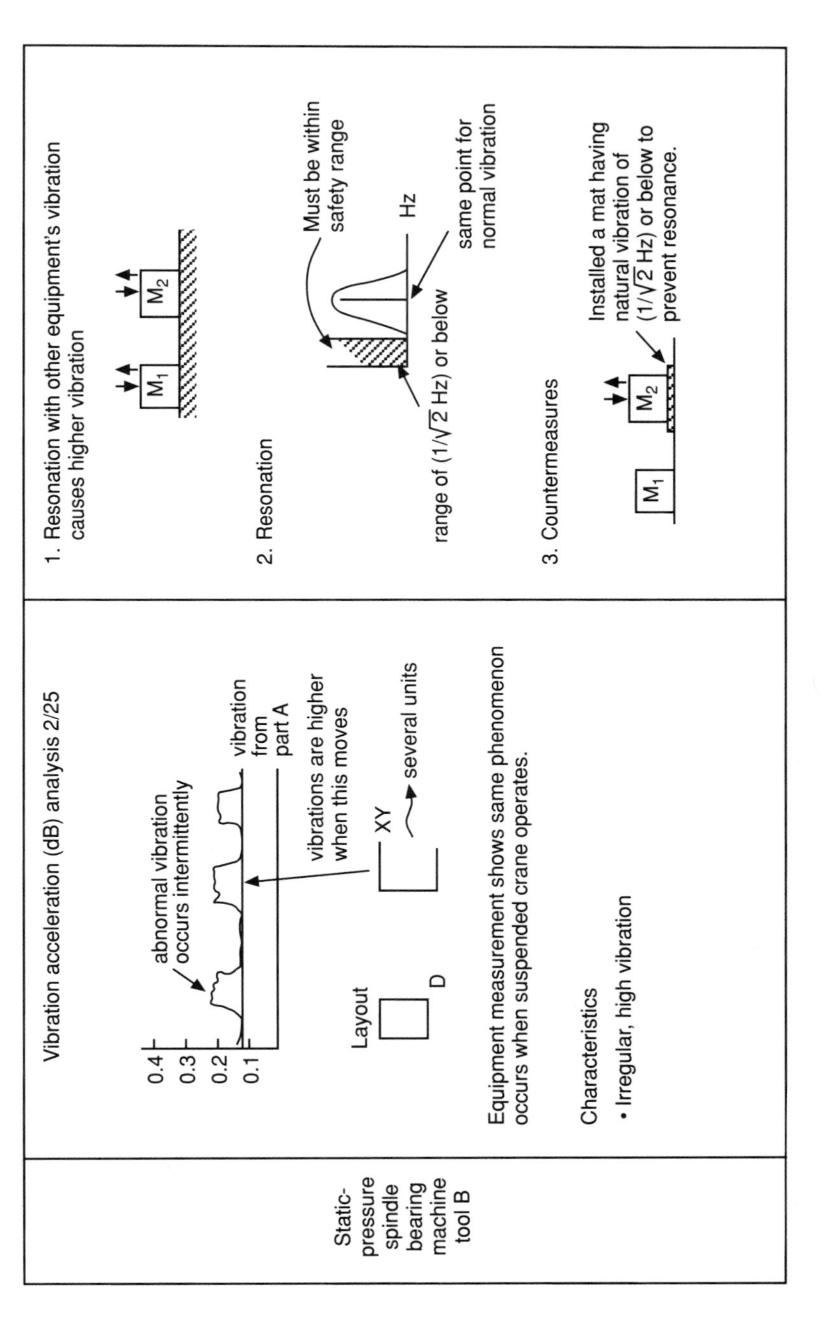

Figure 6-31. Examples of Equipment Diagnostics

7
Standardization for Operation Management

PROBLEM-CONSCIOUSNESS AND THE PIONEERS OF OPERATION ANALYSIS

In the context of this chapter's discussion, "operations" are defined as the work people do as they use materials or machines. All such operations have a sequence; the more waste in the sequence, the slower the operations progress and the more human labor (employee-hours) is required. This creates additional waste in the form of more work for the personnel department.

On the other hand, when the operation sequence includes little waste, the time needed for a certain operation (i.e., the "operation time") is consistent and therefore promotes highly reliable and smooth production scheduling and personnel scheduling. This principle is reflected in the expression, "time is the shadow of motion" — the operating time is a function of the sequence of operations (see Figure 7-1).

Key studies of operation time and methods to improve work efficiency began in the 1920s and resulted in methods used in industry. Let us briefly review the work of some pioneering researchers in the field of operation standards that laid the foundations for standardization of work operations in companies today.

The Work of Taylor

Frederick W. Taylor was the developer of a scientific method of work management. Although "Taylorism" has sometimes been criticized as the pursuit of

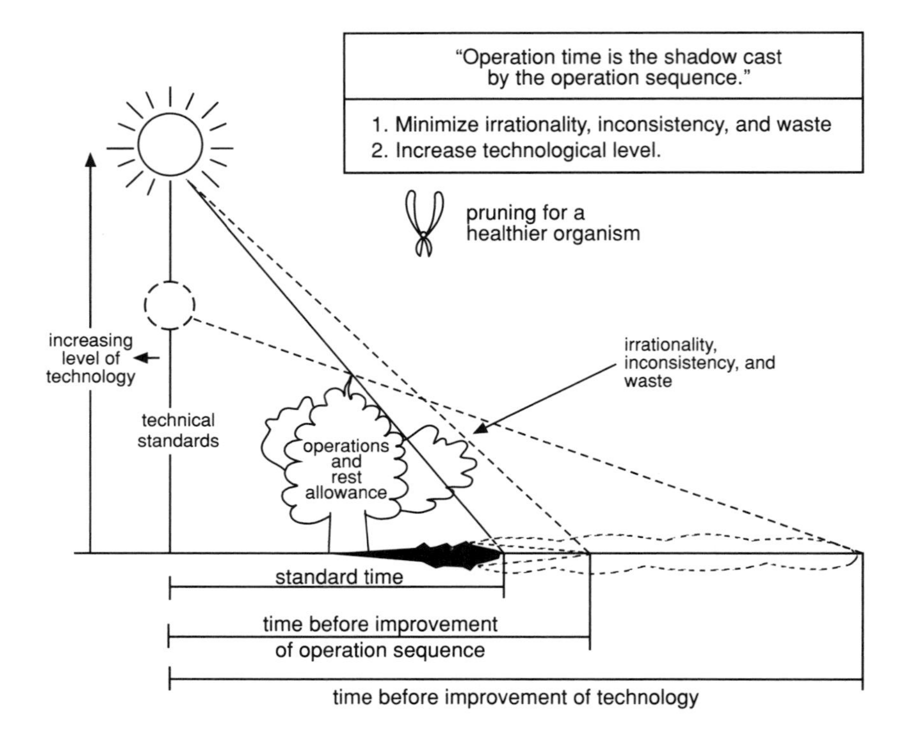

Figure 7-1. Relation between Operations and Operation Time

labor efficiency with total disregard for humanity, Taylor's scientific methods for studying the cutting of metals and other industrial operations made major breakthroughs that earned him respect as a pioneer in the field of industrial management. The standard time system Taylor developed has been modernized with a view toward human needs, and it remains a very important tool.

Taylor's book, *The Principles of Scientific Management,** describes how he developed his standard time system. In the 1880s, Taylor was the general foreman of a steel plant belonging to the Midvale Steel Company. There he began his work in overcoming the perception of "laziness" in work operations. The basic management approach of the era was the carrot-and-stick technique. Managers would offer pay raises for more productive workers who had a greater output. But soon after each increase in output, the management reduced the standard time. Under this system, eventually, higher productivity

* F.W. Taylor, *The Principles of Scientific Management* (New York: Harper & Brothers, 1911).

no longer reaped financial rewards. The shorter standard times made the work more difficult, and workers who could not keep up were penalized with wage cuts. Such problems as unreliable materials supply and low productivity in changeover operations — problems for which the line workers were not responsible — nonetheless caused wage reductions. Ultimately, the workers saw no point in increasing their productivity and lost any desire to do so, creating the problem characterized as "laziness." As a plant foreman, Taylor's challenge was to deal with all of these morale-depressing problems.

Taylor decided to follow the carrot-and-stick approach, but to avoid the problems arising from lack of respect for workers, he wanted to introduce standards for defining a fair day's work. This led him to develop scientific management methods, a study sequence that began with observation and included analysis, generalization, and evaluation to reach his objective of determining standards. Taylor believed that his workers wanted to do good work and would welcome a standard that told them the details of the work sequence and helped them evaluate their own work.

The most important tool for Taylor in his scientific management studies was the stopwatch. He used time as the unit of measurement for evaluating and studying work operations. Interestingly, for his "time study" approach, Taylor found inspiration in the works of Leonardo da Vinci. Shown below are results from Leonardo's time study of shoveling work. (He used a metronome as his clock; each beat stood for one unit of time.)

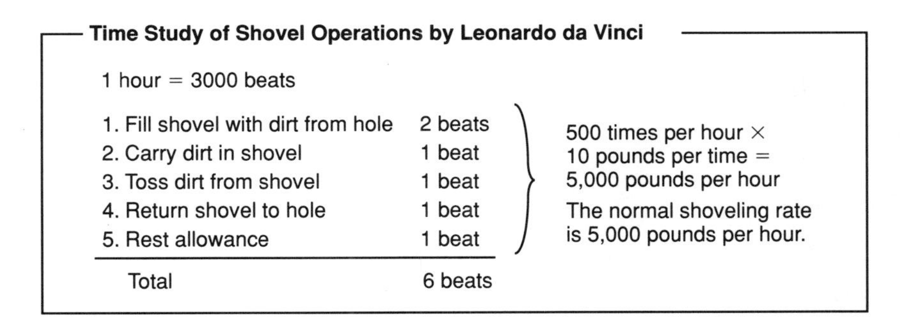

Time Study of Shovel Operations by Leonardo da Vinci

1 hour = 3000 beats

1. Fill shovel with dirt from hole	2 beats
2. Carry dirt in shovel	1 beat
3. Toss dirt from shovel	1 beat
4. Return shovel to hole	1 beat
5. Rest allowance	1 beat
Total	6 beats

500 times per hour ×
10 pounds per time =
5,000 pounds per hour

The normal shoveling rate is 5,000 pounds per hour.

When clearly scientific methods are used to determine operation methods and operation time, it is easier to get workers and managers to agree on what constitutes fair operation standards. Taylor explored these studies even more deeply after he left Midvale Steel Company and began working as a consultant. At one of his clients, Bethlehem Steel, his research in improving operations in the iron ore shoveling yard has become a classic.

The work force in the iron ore yard ranged from 400 to 600 workers. Taylor selected two or three of the more experienced shovelers to train as subjects in his research. The following are some of the results of this research.

1. Taylor determined that each worker could produce the maximum output per day when the shovel load was between 21 and 22 pounds. Heavier shovel loads made the workers more productive at first, but they later tired and produced a smaller volume per day. Lighter loads enabled the shovelers to work faster without becoming too tired, but their total daily output was still lower.
2. Taylor also experimented with shovel shapes to find the best shapes for shoveling sand, stones, or coal. He came up with ten types of shovels, each best suited to a different material.
3. He built a tool room, from which to distribute various types of shovels in accordance with the work instructions.
4. He started a planning department so work operations could be planned in advance and submitted to the foreman.
5. The foreman weighed each worker's output per day and awarded bonuses to workers who exceeded a certain amount. Workers who failed to meet the standard were assigned to a training group to receive new training in standard operations. Those who received this training but still failed to meet the standard daily output were assigned to other work.

The long-term results of Taylor's research were as follows.

No. of shovelers:	400 to 600	→	140
Average daily output:	16 tons per person	→	59 tons per person
Average daily wage:	$1.15 per person	→	$1.88 per person
Labor costs:	$.072 per ton	→	$.033 per ton
Cost savings:	$78,000 per year		

Figure 7-2 shows a model for a system based on Taylor's approach. Today, even computer-based factory management systems use these fundamental scientific management principles.

The Work of the Gilbreths

Frank B. Gilbreth and his wife, Lillian M. Gilbreth, two other renowned pioneers of production management, lived at about the same time as F.W. Taylor.

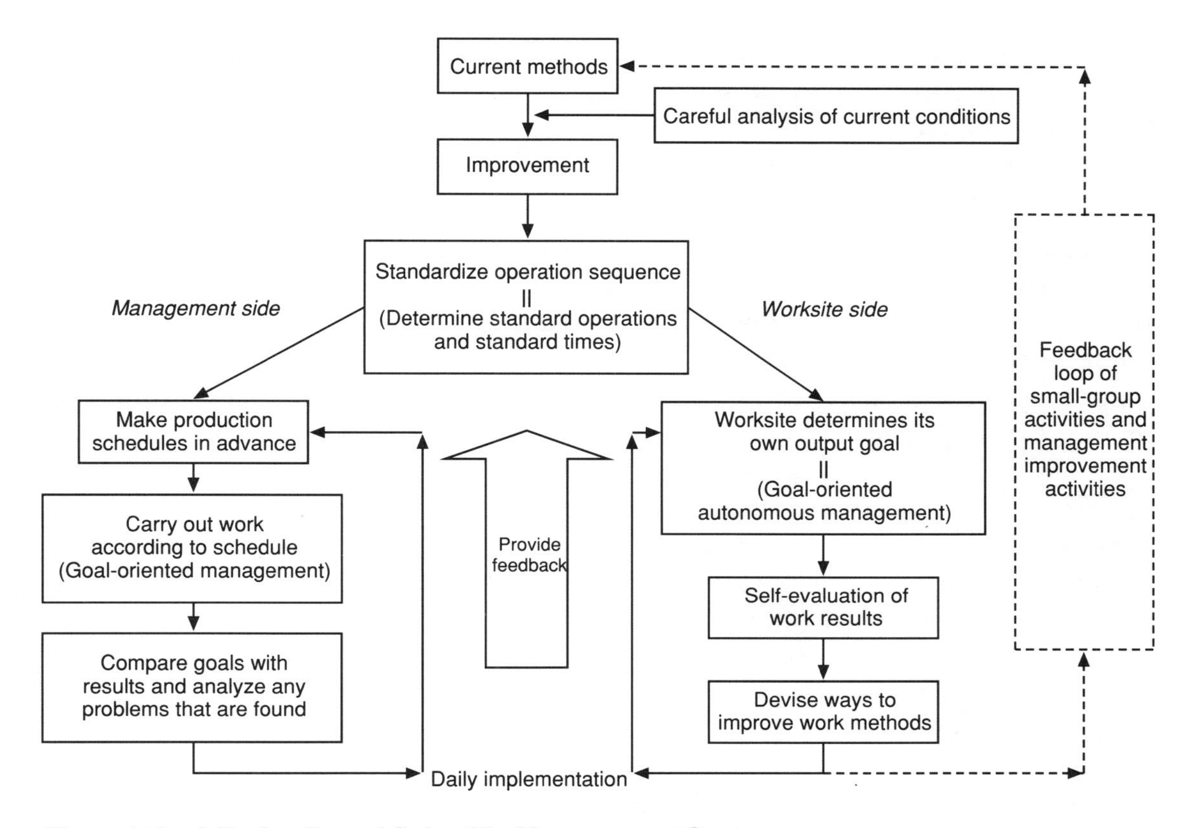

Figure 7-2. A Taylor-Based Scientific Management System

The Gilbreths' most famous achievement was the invention of motion study symbols called *therbligs*, an anagram of their last name. Together they worked for standardization that respected workers' humanity.

The Gilbreths differed from Taylor in that their standardization work centered on the goal of finding the "one best way" to do the task at hand.

At age 17, Frank Gilbreth joined a construction company as an apprentice bricklayer. Bricklaying is an old and widespread trade, practiced by millions over the centuries; however, it is also a trade that had seen very little change in the methods it used.

Gilbreth noticed that each bricklayer in his crew used some techniques he had worked out on his own; no two bricklayers used exactly the same methods. He also observed that the same worker did not always use the same bricklaying motions and that he used three different sets of motions when working slowly or quickly, or when teaching bricklaying to someone else. Gilbreth felt that these differences pointed to a serious problem.

Gilbreth suspected that there must be one best way to lay brick. He thus began making detailed observations and analyses of motions, an approach he later termed *motion study*.

By experimenting with different arrangements of positions for the brick pile, mortar box, and bricklayer, Gilbreth worked out an efficient arrangement that reduced the total number of motions used in picking up brick and mortar and laying the brick from 18 motions to just 5. This big improvement resulted from various small improvements: For example, he designed a scaffold that could easily be set for different heights and had the bricks and mortar box placed on a raised shelf so that the bricklayer did not have to waste time and energy bending down to pick up materials. As a result, Gilbreth nearly tripled the average bricklayer's output, from 120 bricks per hour to 350.

Such dramatic improvement always indicates that the previous methods contained large amounts of waste. Gilbreth's motion study developed faster, easier, and more reliable bricklaying methods. The obstacles he faced in motion study were always forms of waste. For example, he later observed that concrete workers had to stop intermittently to remix the concrete, so he invented a concrete mixer to eliminate this task. After leaving his bricklayer's job, he went on to apply his motion study techniques in various other fields.

Figure 7-3 lists 18 therbligs, which describe basic hand motions; they are classified into three types, relative to the value of the work they perform. These symbols not only simplify description of hand motions, they also facilitate improvements in the sequence of hand motions. This approach is a great way to begin making improvements as part of standardization. Using easily identifiable symbols makes motion study more visual, and visibility is one of the cardinal rules of standardization.

Gilbreth also invented the cyclegraph and chronocyclegraph, two devices that photograph the paths of small light bulbs attached to the hands of an operator as he or she performs a task. This finely detailed observation method is known as micromotion study. The work of the Gilbreths covered a wide range of fields, including the development of process analysis symbols for describing the motions of operators, materials, and machines in production processes.

From the Ford System to the Toyota Production System

Beginning around 1910, Henry Ford and his staff instituted some of Taylor's time study and the Gilbreths' motion study methods on a conveyor line

Type	Name of Motion	Letters/ Symbol	Description or Pictographic Meaning
Type 1: Motions required for performing an operation	1. Transport empty	TE	Empty palm
	2. Grasp	G	Hand open for grasping
	3. Transport loaded (carry)	TL	Hand carrying something
	4. Position	P	Object being placed by hand
	5. Disassemble	DA	Part of assembly removed
	6. Use	U	Letter "U" from the word "Use"
	7. Assemble	A	Several things put together
	8. Release load	RL	Dropping content from hand
	9. Inspect	I	Shape of a magnifying lens
Type 2: Motions that tend to slow down Type 1 motion	10. Search	Sh	Eye turned to look
	11. Find	F	Eye finding object
	12. Select	St	Pointing to object
	13. Plan	Pn	Person thinking
	14. Pre-position (setup)	PP	Bowling pin ("set up")
Type 3: Motions that do not perform an operation	15. Hold	H	Magnet holding a bar
	16. Unavoidable delay	UD	Person falling accidentally
	17. Avoidable delay (standby)	AD	Person lying down voluntarily
	18. Rest	R	Person seated

Source: Kenichiro Kato, *IE for the Shop Floor: Productivity Through Motion Study* (Cambridge, Mass.: Productivity Press, 1991), 61.

Figure 7-3. Gilbreth Therbligs

designed for a standardized production flow. Ford's production-line experiments yielded many successful results, including faster machining of parts, development of specialized production equipment, standardized operations that eliminated the need for experienced workers, job assignment specialization, and production flow organization. These developments are known as the Ford mass-production system, or "the Ford System."

Henry Ford got some of his ideas about production flow by observing operations at a meat-packing plant. Seeing sides of beef being sent smoothly down a long table, with sections of meat separated from bones at various stages along the way, it occurred to Ford that instead of taking something apart, he could reverse the flow to put things (automobiles) together.

These improvements enabled Ford to lower the price of his cars from $850 to $600; true prosperity, he believed, is marked by reduction of prices. His major accomplishments included

- Improving quality and unit casting methods at a foundry
- Developing a multispindle drill press that can process four sides of a cylinder block
- Developing more rational layout of people, machines, and materials in assembly plants and using conveyors for parts

Thanks to these innovations, the Ford system's productivity rose dramatically: By 1913, the chassis assembly time per vehicle had been reduced from twelve hours to just one hour. In the 1920s, the Ford Motor Company's development of the transfer machine, a multi-item production machine for automobile parts ushered in a whole new era of production technology.

The Ford system was the original model for today's automotive processing and assembly lines. However, this conventional model contains many elements that are ill-suited for today's era of wide-variety, small-lot production and rapid changes (such as changes in output volume and the trend toward shorter product life cycles).

Today's markets require inexpensive one-piece flow production. The production system developed at Toyota Motor Company enables factories to meet these needs. The average automobile, however, contains more than 10,000 parts. When there are dozens of models to assemble, each made to order, the production line can get very complicated. This has led automakers to do whatever they can to establish just-in-time parts delivery systems. This involves — especially in the context of wide-variety, small-lot production — dealing with the following problems:

1. A large variety of parts naturally leads to large parts inventories. This is especially a problem when the factory inventory also includes cars that have no buyers (i.e., dead inventory).
2. The unavailability of even one part can stop the assembly line. Such stoppages can bring considerable labor and idle equipment costs. Managers also spend a lot of time figuring out where to store supplies of parts needed at various stages. As in-process inventory accumulates, it begins to conceal product defects and equipment abnormalities because of the time lag in using up these parts.
3. Greater product variety tends to require more frequent changeovers, which lowers net productivity and prevents the production line from flowing smoothly.

The Toyota Motor Company came up with a wide-variety, small-lot production system designed to solve these kinds of problems. Toyota's approach is rooted in the elimination of waste from production, the maintenance of minimal work-in-process, and a flexible production line that can handle small lots of various models in response to the current market needs. "Leveled production" is one name for this innovation in production line organization and scheduling. The *kanban* system, a management tool used to convey production and supply information in the Toyota production system, is well known not only in Japan, but internationally as well.

Other books describe the Toyota production system in detail.* For present purposes, consider the summary description of this system in Figure 7-4. Please note that standardization of operations is a fundamental part of the system.

PRACTICAL TECHNIQUES FOR DEVELOPING AND DISPLAYING OPERATION STANDARDS

A company can take various approaches to raise the level of its production technology, including FA (factory automation), FMS (flexible manufacturing systems), and CIM (computer-integrated manufacturing); Figure 5-20 lists these

---

* See, for example, Shigeo Shingo, *A Study of the Toyota Production System from an Industrial Engineering Viewpoint* (Cambridge, Mass.: Productivity Press, 1989); Taiichi Ohno, *Toyota Production System: Beyond Large-Scale Production* (Cambridge, Mass.: Productivity Press, 1988).

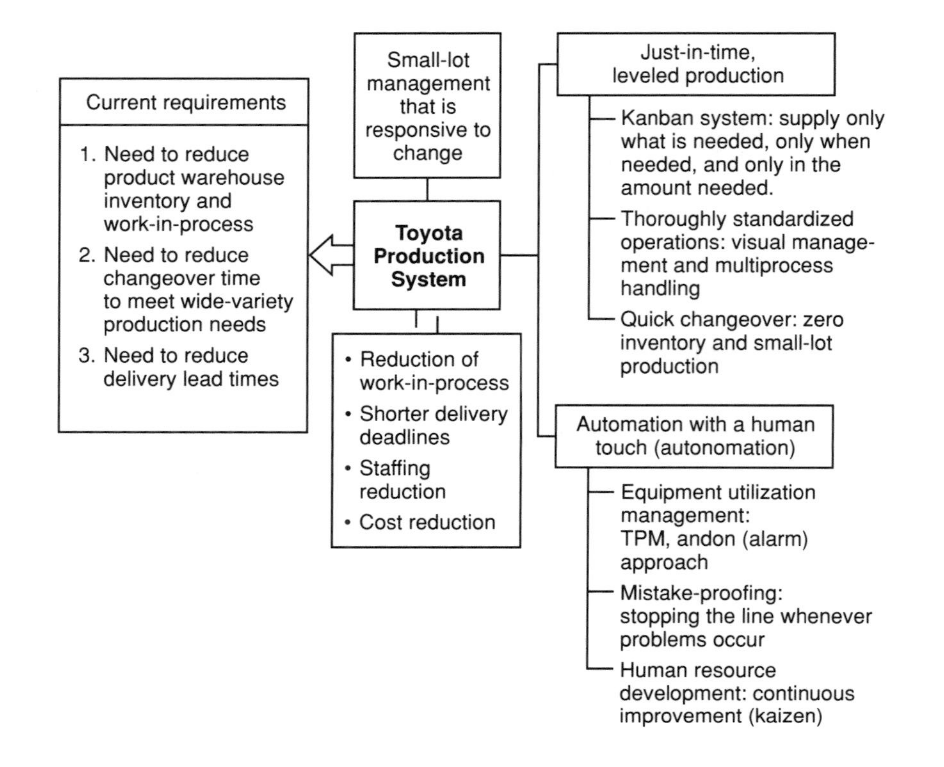

Figure 7-4. Organization of the Toyota Production System

approaches. Even as companies pursue these high-tech approaches, however, they are still well advised to exercise the same problem-consciousness that Taylor and the Gilbreths used so many years ago. Problem-consciousness and a continual determination to solve the problems are common themes from Taylor to the Toyota production system.

As a prerequisite to understanding problem-consciousness, consider the following steps and countermeasures companies can take in standardizing operations to raise the level of production technology and improve the company organization.

Different Methods for Different Levels of Standardization

The starting point begins with determining the methods to use for dealing with operation problems. To do this, you need to look at operations according to the time they require. Any operation that exceeds the ideal time must be viewed as waste. Figure 7-5 shows a time-based analysis of operations. This approach is a valuable way to identify improvement needs.

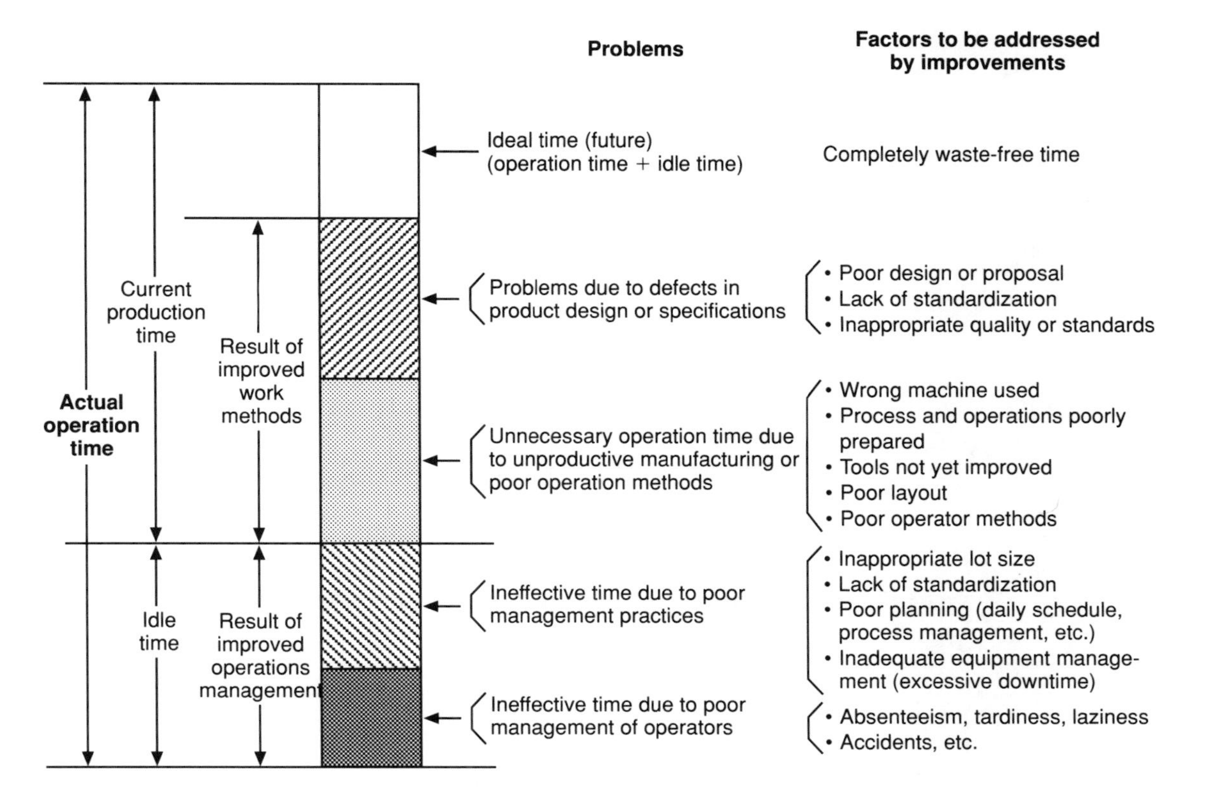

Figure 7-5. Using Operation Times to Identify Operation Problems

The next step actually makes the improvement. Figure 3-4 shows the various levels of improvement making; here we will discuss two more procedures that support those improvement levels.

Figure 7-6 diagrams the progression toward increasingly effective improvement methods (methods that combine existing technologies or introduce new ones). Figure 7-7 looks at the same steps from the perspective of time-based measurement and ideal operation times. As such, Figure 7-7 is an important aid for choosing improvement targets and techniques best suited for each step.

You could use the handle of a screwdriver to hammer a nail, but the best tool to use is a hammer. Likewise, you should also select the most appropriate methods for making improvements, according to the particular target and objective of improvement. Matching improvement methods with improvement needs is one means of standardization. Unfortunately, many companies still use the kinds of methods shown at the bottom of Figure 7-6, superficially treating symptoms without effecting any cures for the problems at hand.

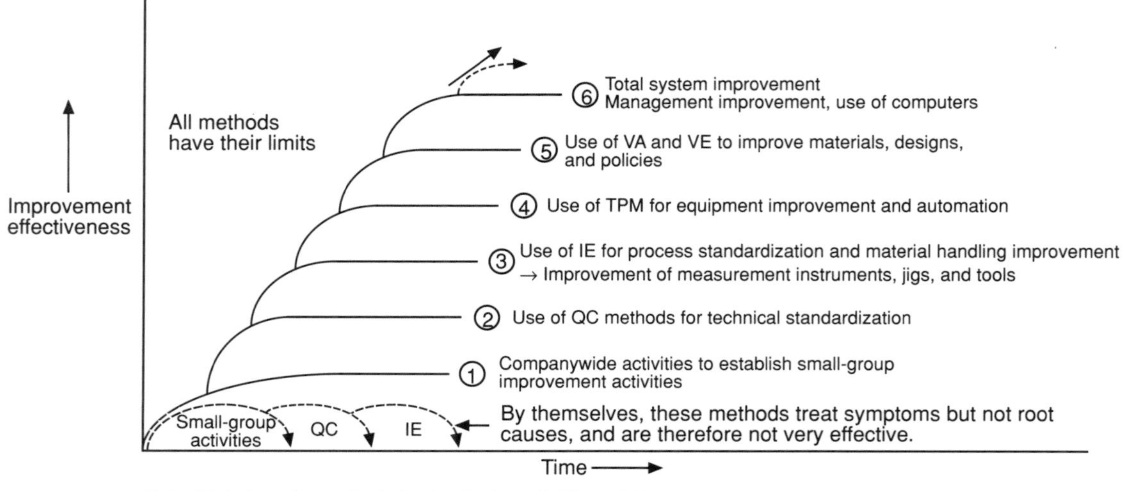

Note: Circled numbers refer to the levels shown in Figure 3-4.

Figure 7-6. Progression of Increasingly Effective Improvement Methods

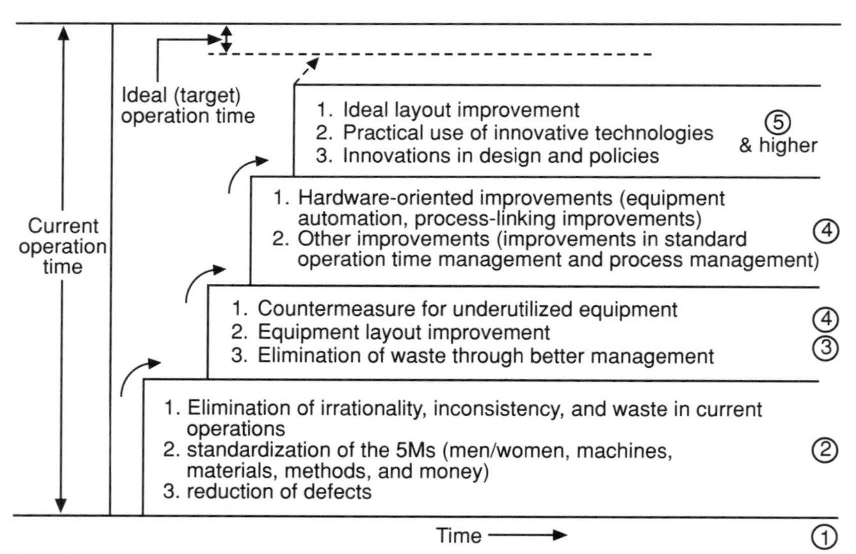

Note: Circled numbers refer to the levels shown in Figure 3-4.

Figure 7-7. Progression of Improvements Toward an Ideal Operation Time

For example, Figure 7-8 shows various methods used for time study. Each method has its own strengths and limitations; as in the screwdriver and hammer analogy, you should take care to select the method best suited to the improvement target and objective.

Steps in Making Operation Standards

The steps taken when making operation standards vary somewhat according to the methods used, but are generally as shown in Figure 7-9. These standardization steps are very similar to the P-D-C (plan-do-check) steps for quality improvement activities. In both cases, it is important to continue repeating the cycle of steps while making improvements.

In the steps up to the point of proposing new methods, you are making improvements in operating methods. In the later steps, you are establishing new operation standards that incorporate these improved methods. From this new operation standard a new standard operation time emerges.

The stopwatch method is one of the most often-used approaches for making new operation standards. Figure 7-10 describes the time study process and

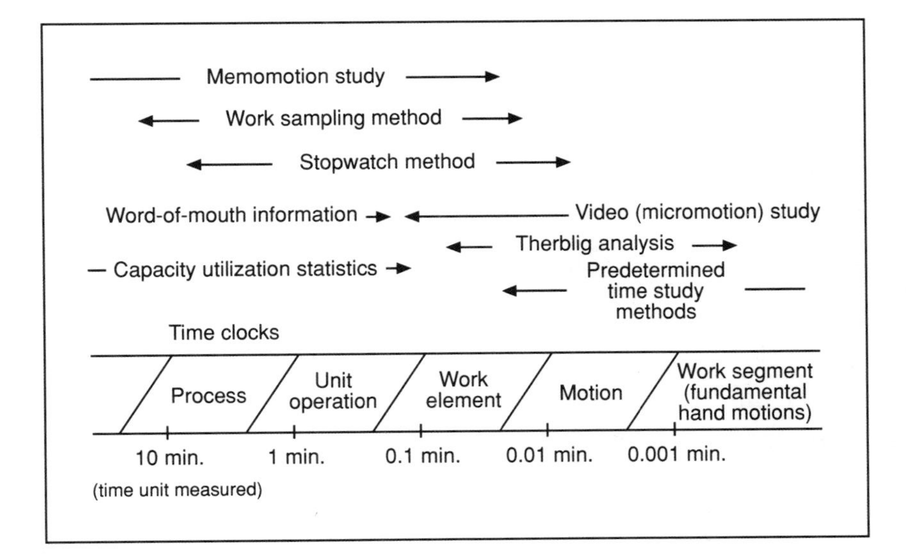

Figure 7-8. Various Time Study Methods

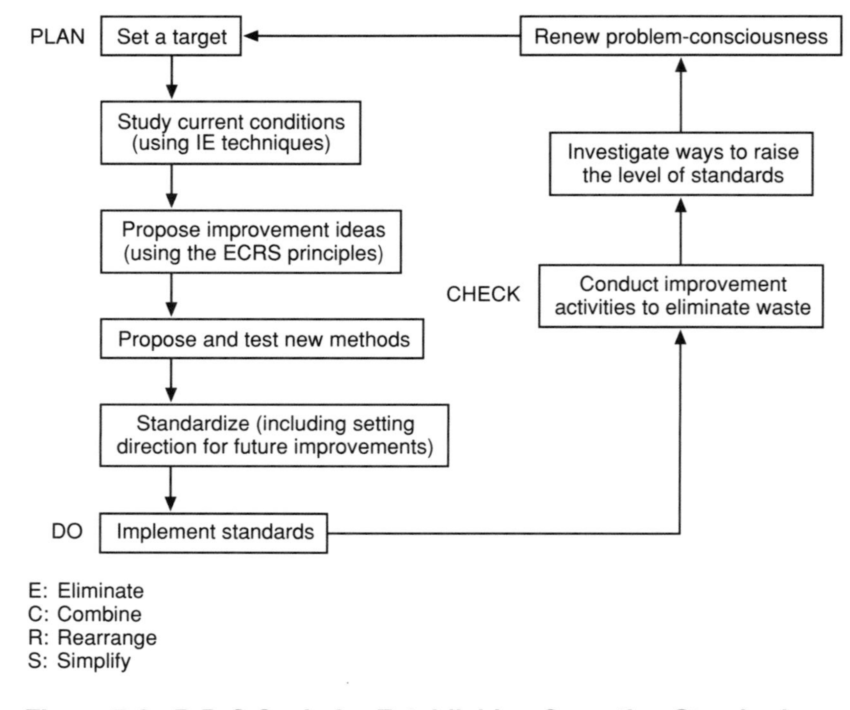

Figure 7-9. P-D-C Cycle for Establishing Operation Standards

steps for establishing operation standards using this method. Figure 7-11 shows an example of a process standard sheet that averages the results of a stopwatch time study for nine sequential operations to determine a standard time for each operation.

Good methods for determining standard operations include:

1. Using a stopwatch to determine standard time values
2. Creating an operator-machine analysis chart
3. Creating a process analysis chart

The subsections that follow describe these methods. It is also helpful to display combinations of operations, as when using the PERT method. Showing the timing of electrically controlled machine movements during the operation may help as well. Select the methods and subjects according to the time study subject, objective, and degree of technical difficulty. In any case, making and maintaining the improvement is more important than the method used to display the improvement results.

1. Operation improvements were made (determine operation requirements, tools, environmental conditions, etc.)
2. We determined the operation sequence (process).
3. We selected the object of measurement and the operator(s).
4. We used a stopwatch to measure the net operation time per product unit (in this case, 2.5 minutes per unit).
5. We used the work sampling method to measure the process over an 8-hour (480-minute) operation period. The following represents non-value-adding time that was measured:
   (1) Random downtime (machine failures, parts outages, etc.; little repetition, but still unavoidable within operation time): 3% (14.4 min.)
   (2) Worksite time allowance (morning meetings, etc.) 3% (14.4 min.)
   (3) Personal time allowance (lavatory time, etc.): 4% (19.2 min.)
   (4) Other avoidable elements (including some that should be improved): 5% (24.0 min.)
6. Work continued smoothly throughout the day, yielding an output of 164 product units.
7. The net operation time was estimated.
   164 units × 2.5 min./unit = 410 min.
8. The fatigue allowance was calculated.
   480 min. − (410 + 28.8 + 19.2) = 22 min.
   Investigating the contents of the 22-minute fatigue allowance, and referring to a relative metabolic rate (RMR) chart, we determined that with improvement, a 20-minute total fatigue allowance was appropriate.
9. We established a new allowance rate:
   (1) Random downtime allowance + worksite time allowance = 28.8 min./410 min. = 7.0%
   (2) Personal time allowance = 19.2 min./410 min. = 4.0%
   (3) Fatigue allowance = 20.0 min./410 min. = 5.4%
   TOTAL = 16.4%
10. We established the standard time (ST):
    ST = 2.5 × (1 + 0.164) = 2.91 min. per unit
11. We displayed our analysis results and wrote them up as an operation standard sheet for use in operator training.
12. We also made visual management display boards showing these results as process standards.

Figure 7-10. Operation Analysis and Establishment of New Operation Standard via the Stopwatch Method

Displaying Standards Determined by the Stopwatch Method

Figure 7-12 provides an example of an operation-standard display method used often for changeover operations. This type of display makes the important points regarding both the technical and the process standards easy to see; it is

Company: Company N

Product code:	SS
Diagram No.:	A-36
Material:	FCMB32
Process:	Cast finishing
Equipment:	11-kw grinder/polisher
Grinding speed:	1000 km/minute
Regular operator:	TY
Observation date:	1/25

Diagram

d.
i.
a.
g.
b.
c.
h
e.
f.

Standard: Must be ground even with surface of casting.

No.	Work element	1	2	3	4	5	6	7	8	9	10	Total	Average	Rated value	Net time
1	Pinion boss a. Grind front of blower hole	34	35	34	36	36	37	35	35	36	37	355	36		
		0:34	80	3:27	78	6:28	85	9:32	81	12:33	87				
2	b. Grind oil inlet	14	14	15	14	15	15	14	16	15	14	146	15		
		48	94	42	92	43	8:00	46	97	48	14:01				
3	c. Grind periphery of casting	5	5	4	5	5	5	4	6	5	5	49	5		
		53	99	46	97	48	05	50	11:03	53	06				
4	d. Grind oil inlet	15	16	16	15	17	16	17	16	16	16	160	16		
		68	2:15	62	5:12	55	21	67	19	69	22				
5	Base e. Grind burrs	18	18	19	20	18	18	20	19	19	20	189	19		
		86	33	81	32	83	39	87	38	88	42				
6	Peripheral base f. Grind burrs	14	12	11	11	12	11	11	10	12	13	117	12		
		1:00	45	92	43	95	50	98	48	13:00	55				
7	Four feet on upper side g. Grind burrs	20	21	22	20	24	20	21	22	22	20	212	21		
		20	66	4:14	63	7:19	70	10:19	70	22	75				
8	Slanted surface next to feet and in-row fitting	12	15	14	15	14	13	13	12	13	13	134	13		
		32	81	28	78	33	83	32	82	35	88				
9	Opposite side i. Grind burrs.	13	12	14	14	15	14	14	15	15	14	140	14		
		45	93	42	92	48	97	46	97	50	15:02				
10													1:51	105%	1:58

Comments

Grindstone: A16NB Size (mm): 610 × 50 × 152.4
Operator: 4 years' experience, high skill level
Additional time was needed for grindstone dressing and changeover
Safety regulations: Protective clothing and safety goggles required

Basic time:	1:58 minutes
Misc. time:	0:05 minutes
Total time:	1:63 minutes
Time allowance coefficient:	1.164
ST:	2:00 minutes

Note: The upper figures in each row are the observed times (calculated values), measured in decimal seconds. The numbers in the lower, shaded rows are the cumulative time on the stopwatch. Cumulative time is measured in sequence from workpiece 1 through 10; for simplification, the digits for full cumulative minutes appear only once for each minute.

Figure 7-11. Example of Operation Standard Sheet (Process Standard)

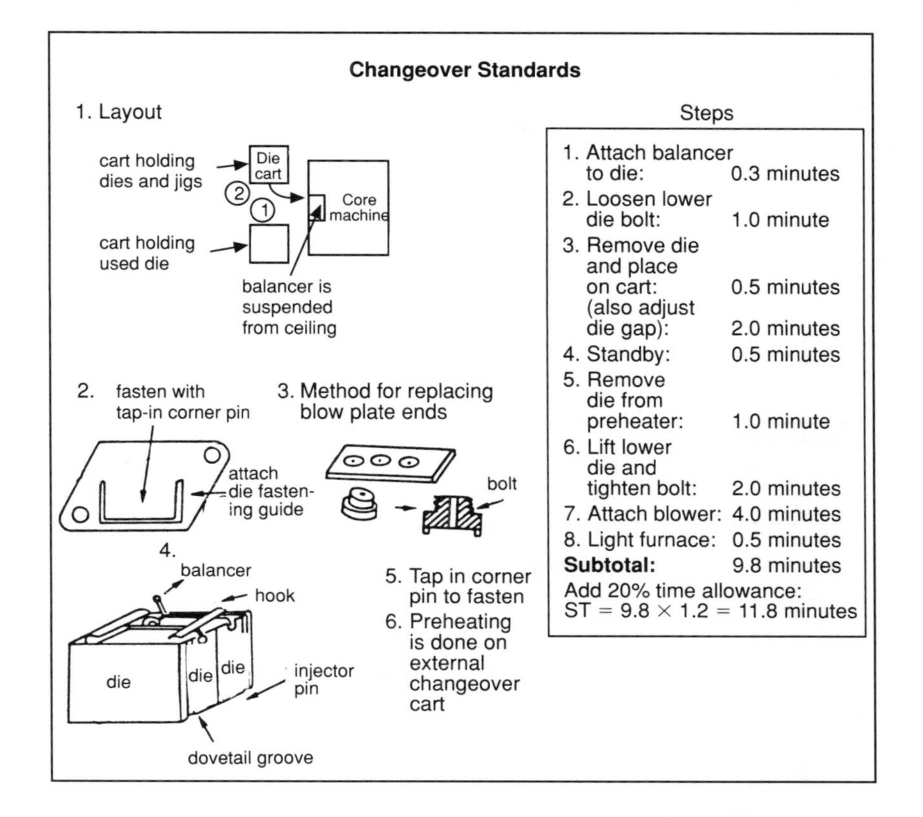

Figure 7-12. Example of Standards Display for Changeover Opertions (Using the Stopwatch Method)

thus a handy reference for operators and work site managers. As a further improvement, a bar graph of the time values could be added to make it easier to see which operations take most time.

Figure 7-13 shows an operation standard sheet of the sequence of hand motions used in detailed work, such as assembling wires. The addition of therblig symbols to help describe the hand movements might make this even more easily understood and help trainees learn the operation sequence faster.

Displaying Standards Using Operator-Machine Charts

Figure 7-14 shows standards that enable a single operator to handle three machines efficiently. These standards result from operator-machine analysis,

Operation: Attach resistor wires	Product: XXX	Parts: A, B, C, and D	Process: Parts setup	Drawing No.: B-64

Operation diagram and processing diagram:
Operator position diagram

R: Reach forward
Gr: Grasp
M: Move (carry)

Machines, tools, and jigs used:	Comments from observer:
Wire laying jig	

Left hand				Total time RU*		Right hand			
No.	Motions	Parts	RU			RU	Parts	Motion	No.
1.	Standby	BD			7	7	C–1	Reach for resistor	1
2.	"	"			12	5	3–V	Grasp resistor	2
3.	"	"			18	6	B–2	Move resistor	3
4.	"	"	25	25	25	7	CT > 10 r > 09	Set resistor in jig	4
5.	Reach for wire	A–2	4	29			BD	Grasp resistor	5
6.	Grasp wire		0	29			"	"	6
7.	Bend and Move wire	A–2	4	33			"	"	7
8.	Release wire		0	33	33	8	"	"	8
9.	Standby	BD			38	5	B–1	Move resistor	9
0.	"	"	6	39	39	1	O–	Move resistor	0

Analyst	Supervisor	Approval		Total measured time	Time allowance coefficient	Standard time
				.039	1.25	.05

*RU is the time unit used in Ready Work-Factor analysis.

Figure 7-13. Operation Standard Sheet for Assembly of Small Items (Using RWF Method)

which studies how work is divided between operators and machines in terms of the respective operation time values for the operator and the three machines. This chart is a useful tool when looking for the least wasteful combination of operations to meet the needs of different product models that have different

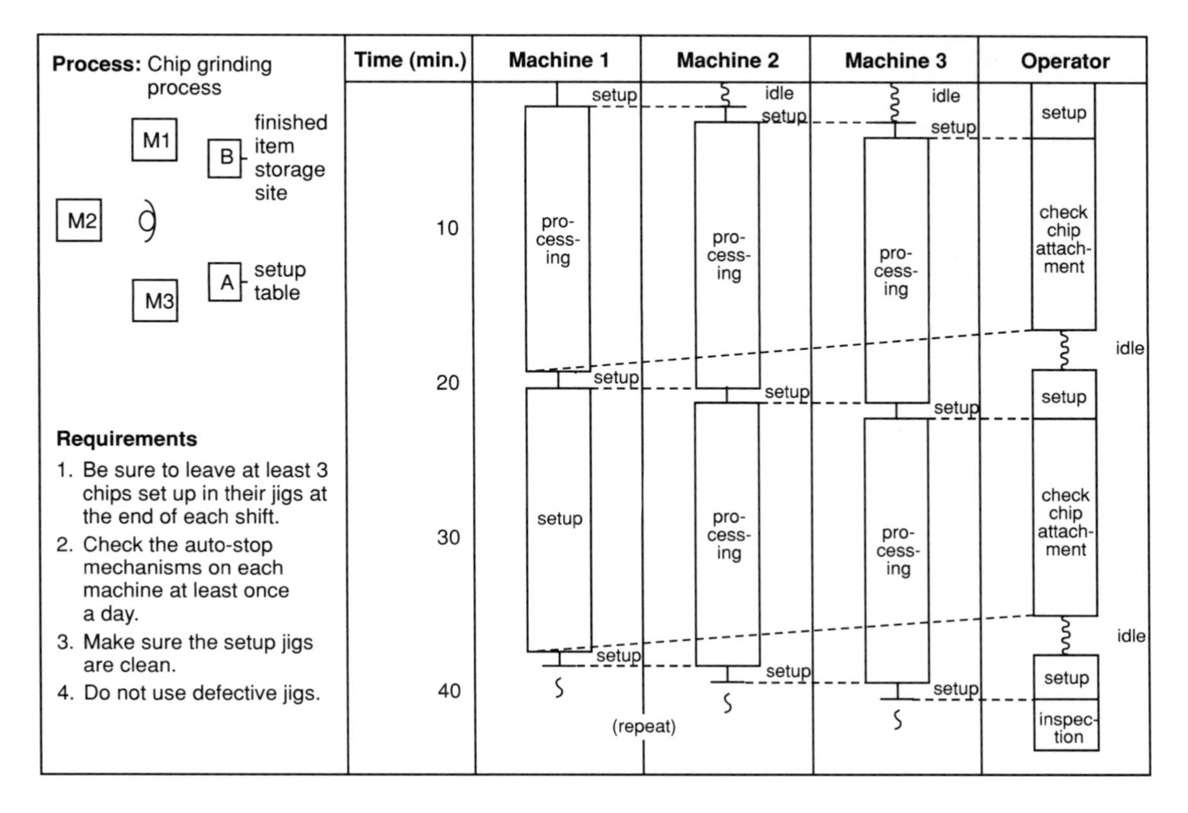

Figure 7-14. Operation Standard Sheet Using Operator-Machine Analysis

cycle times. By including the operator's motions in the chart, you can use the chart to show standards for multiprocess handling, described later.

Displaying Standards Using Process Charts

Figure 7-15 shows another operation standard sheet, this time using process analysis to make it easy to understand the sequence of operations at an assembly process. The process analysis symbols in this chart (explained in Figure 7-16) help operators view the operation from a fresh perspective and also help make waste easier to identify.

No matter which kind of operation standard sheet you use, the goal is to make it easy to understand. Many methods are available; choose an approach that makes the key points of the operation clear to the operator. For example,

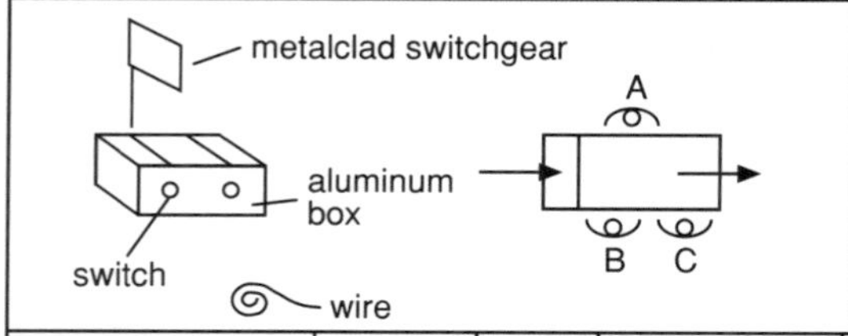

Serial No.: A-800	Lot size:	200
Model: Model B	Drawing No.:	00348
Operation: Metalclad switchgear assembly	Material:	Aluminum
Operator: N	Standard rev. date:	4/10/91

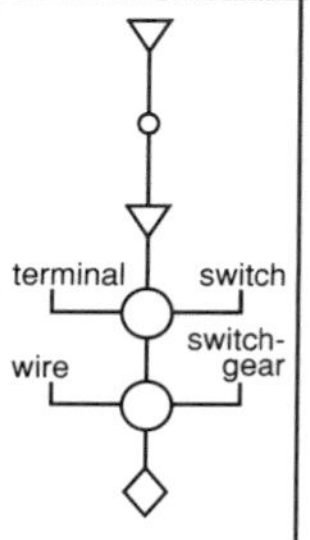

Material handing, etc.	Lot	Time	Process	Operation	Operator	Operation points
Remove from warehouse				Storage shelf management	Wires	Place in specific containers. Attach kanban.
Conveyance cart	20 units			Transport to work table	Wires	Do not stack boxes. Do not remove metalclad switchgears from plastic bags.
Store in boxes	"		terminal, switch	Standby at work table	Wires	
Manual feed	1	100	wire, switch-gear	Assemble	A	Use colors as shown on diagram
"	"	120		Assemble	B	Use test to set after checking wire
"	"	100		Inspect	C	Check using program

Figure 7-15. Operation Standard Sheet Using Process Analysis

Symbol	Meaning	Comments
○	Operation	Check for 100% efficiency in operation speed and contents of operations.
⇨ or ○	Transport, conveyance	Transport does not add value to the product and should therefore be eliminated whenever possible. Items that must be moved should be moved along wide paths and across short distances.
□	Inspection	A square within a circle indicates an inspection process that builds quality into products. Consider ways to increase the precision of inspections, such as by installing mistake-proofing devices (poka-yoke). When this cannot be done, try to establish rapid feedback regarding defects.
△	Delay, standby	Such nonproductive time should be minimized. Bottlenecks result from poor process balancing, poor conveyance methods, inappropriate lot organization, poor production scheduling, and so on. Make corrective improvements.

Figure 7-16. Process Analysis Symbols

see the way Figure 7-17 displays the work sequence and standard times for a multiprocess handling operation. Other methods that facilitate process scheduling management include PERT charts and Gantt charts (described later).

PROBLEM-CONSCIOUSNESS COMES BEFORE METHODS

The following examples underscore the importance of problem-consciousness in standardization.

Improving Without Improvement Methods

A student in an IE course I taught, who worked for a car telephone service company, related this example. Because it was an elementary course in IE methods, this student, *A*, did not have a very extensive grasp of IE when he returned to his company to instruct others in IE methods. Consequently he ran into some difficulties.

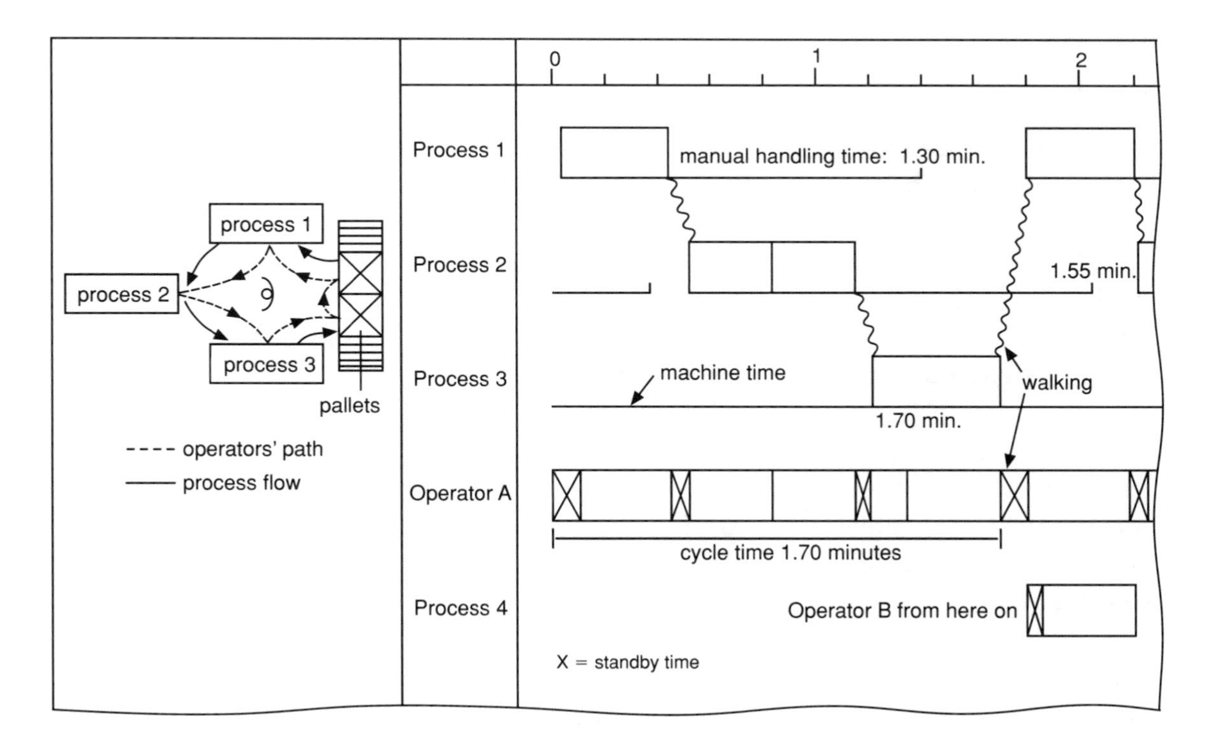

Figure 7-17. A Multiprocess Handling Operation and Standard Time Chart

A reported to me that he had recorded a day's operations with a stopwatch, but was unable to apply his time measurements using a predetermined time system approach. He wondered if he could use the work sampling method instead. I said to him, "The most important activity in IE is not determining which specialized method to use. The basic element of IE is problem consciousness. Why don't you begin by identifying and quantifying the problems?" Before I left, he described various aspects of the operations at his company.

On another day, *A* reported that he was having problems again. His company wanted to reduce the amount of overtime in the car phone service operations and, if possible, to free up some car phone service workers for other kinds of jobs. Our conversation went something like this:

A: You mentioned the other day that service industries face various kinds of problems. In my company's case, the most difficult problems seem to involve installing car phones. I would like to know how I should investigate these problems.

N: Well, perhaps you should start by checking into exactly where and how these car telephones are installed.

A: Okay, I'll try it.

Then *A* left, and I did not see him until a month later.

A: It worked this time! First, I asked the car phone installation person where and how he installed the phone, and he just said "there are lots of ways — it depends on the car." So I checked into it myself.

N: And what did you find out?

A: The car phone can be installed in any of eight different locations, but the three most common locations account for about 80 percent of the installation jobs. This gave us a priority to examine further.

N: So, what did you do next?

A: I visited customers who had their car phones installed in one of the five less-common locations. When I asked them why they chose their particular location, most said they had no particular reason and decided almost arbitrarily.

N: I see. You came upon an important point that got you ready to begin standardizing.

A: Yes, it did. I learned that the three most common locations were the important ones and that the other five positions had no good reason to exist. Next, I got a stopwatch and timed the installation of car phones in the three important locations. I then tried to find ways to improve the installation procedures, with better-organized parts and tools and instructions that anyone can follow reliably, easily, and quickly. I was very happy when I drastically reduced the time we keep our customers waiting while installing their car phones. And I was even happier to find that beginning this week, we've reduced our car phone installation crew from 20 people to just 10, and we're now reassigning the other 10 people to new types of service jobs. I'm sure my company has already recovered the cost of sending us to the IE course.

This example was also instructive for me. As an example of the work sampling method, it did not primarily involve use of the stopwatch method, but *A*'s problem-consciousness led him to use the principles of both of these methods. This demonstrates the fundamental nature of problem-consciousness; specific methods are secondary by comparison.

A 30 Percent Rise in Administrative Department Efficiency

This case concerns an administrative department that was shorthanded when one of its employees left the company. The department was unable to fill the vacancy right away and had to try to make up for the shortage with their remaining staff. No one, however, had a good grasp of the departed worker's tasks. Finally, the group decided to do a time study to figure out how to best perform the work.

Because this was an administrative department, the clerical workers often received assignments from various managers. During particularly busy times they had trouble keeping up with the work load, even though they put in lots of overtime. Some managers got frustrated when the work they handed out came back late. They saw the staffers doing their best and putting in long hours, but they were still unhappy that the work wasn't ready.

For the time study, each clerical worker kept a running record of all the work done during the day, each day for one week. The workers compiled and sorted the data to see how much time each kind of job took. They found that no more than 3 percent of the workday was spent on breaks, but they also found lots of clerical bottlenecks, peak demand periods, and other big problems. Most of the problems stemmed from tasks that were the responsibility of the managers, not the clerical workers.

The department met to study and discuss the data, and decided the following:

1. The clerical processing system needed revision.
2. The department needed more photocopy machines to reduce bottlenecks. Managers who needed only a few copies made should do the copying themselves. A system that limits the volume per request and smoothes out the demand periods should handle the large copy requests.
3. Employee housekeeping activities, such as dishwashing to prepare for guest visits, should be better organized and scheduled to smooth out demand periods and reduce overtime.

The clerical workers were happy to see that the managers had a better understanding of all the work they did.

In my opinion, these methods all needed standardizing. Later, employees in another department requested my help in trying to increase their efficiency by 30 percent. This department had about 100 employees. I decided to use a problem-solving matrix approach, which is a Japanese improvement method of having employees and their managers brainstorm improvement ideas for various processes and improvement keys. I asked each employee to fill out a self-recorded time study data sheet, such as the one shown in Figure 7-18. The employees found the self-recording task fairly simple and consequently the time study went quite smoothly.

In addition, every day each administrative employee filled out an individual data sheet, such as shown in Figure 7-19. From this came a large number of improvement ideas. The employees also made a flowchart of their work and of the standardization of the improved work flow (see Figure 7-20). These improvements enabled one department to absorb operations that had been spread over several departments. In short, they managed to achieve their goal of a 30 percent rise in efficiency.

These two examples demonstrate that the dual activities of thinking up improvements while making analyses, and standardizing while making improvements are merely implementing the steps described in Figure 7-9. No matter what improvement target or work site conditions you are dealing with, it is still best to follow these same standardization steps.

I would caution against skipping any of the basic steps or rushing to introduce computerization, for I have seen many failures result from such hasty measures.

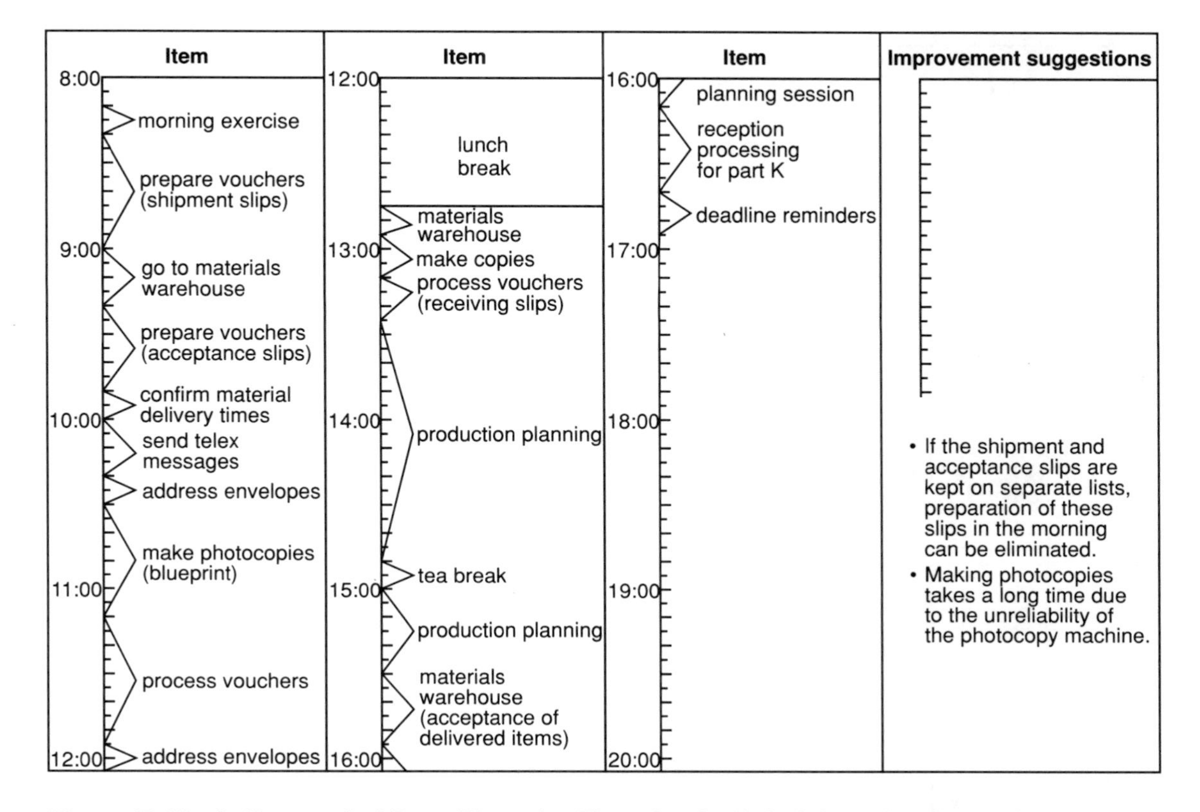

Figure 7-18. Self-recorded Data Sheet for Time Study (Administrative Operations)

The key is to keep repeating the same cycle of basic steps: study the current work-site conditions, make analyses, devise and implement improvements, and then standardize the improvements. After you repeat this cycle a few times, your office or factory will reach a level of standardization where computerization comes more easily and reliably. Administrative departments tend to have the least experience with this type of improvement and standardization and, therefore, are the most likely to leap prematurely toward computerization. They would do well to remember — "haste makes waste."

DISCOVERING THE ADVANTAGES OF A STANDARD TIME SYSTEM

After establishing standard times, you will use them to make a production schedule and carry out production management activities. The foundation for these activities includes the scientific management methods pioneered by Frederick Taylor (see Figure 7-2). With the recent trend toward computerization,

- Please enter data into this sheet every day.
- Please list any improvement ideas you get from this time study.
- You may design and use your own data sheet if you wish.

(Unit: min.)

Item / Month/day	5/13	5/14	5/15	5/16	5/17	Total	Improvement ideas
Morning exercise	20	20/40	10/50		20/40	70	1. Standardization of operations and assignment of duties to shop-floor operators • Process management • Issue order slips (for currently used tools, etc.) and warehouse management • Buff polishing requests 2. Rod lathework arrangements and assignment of duties
Organize documents	20			15/35		35	
Make delivery deadline reminders by phone	15	15/30				30	
Design and drafting for jigs and tools	295		475/770	515/1,285	310/1,595	1,595	
Manufacturing meeting (with tool manufacturer)	50					50	
Blueprints (drawings)	5					5	
Equipment improvements (planning, design, drawings)	85	235/320				320	
Jig and tool manufacturing meeting (with manufacturer)	40		70/110	10/120	50/170	170	
Process follow-up (meeting)	10	35/45	20/65			65	
Issue order slips	5	30/35	10/45		150/195	195	
Check up on production conditions	10					10	
Review follow-up	10	20/30				30	
Arrange for required materials (confirm delivery deadline)	45					45	
Fill out trip report		75				75	
Order required jigs and tools		55				55	
Receive ordered Hi-TORK JIK		195				195	
Fill out warehouse receptions instructions		10				10	
Study product defect statistics			10	5/15	135/150	150	
Help with shaft inspection			40	120/160		160	
Request hobbing			10	5/15		15	
Arrange for defective products to be returned to K				105		105	
Attend English lesson				80		80	
Make other phone calls						15	
Total	610	695/1,305	655/1,960	855/2,815	665/3,480	3,480	

Note: When two time figures are shown divided by a slash, the first figure is the time taken on that day and the second time is the cumulative time for the week.

Figure 7-19. Individual Time Study Data Sheet

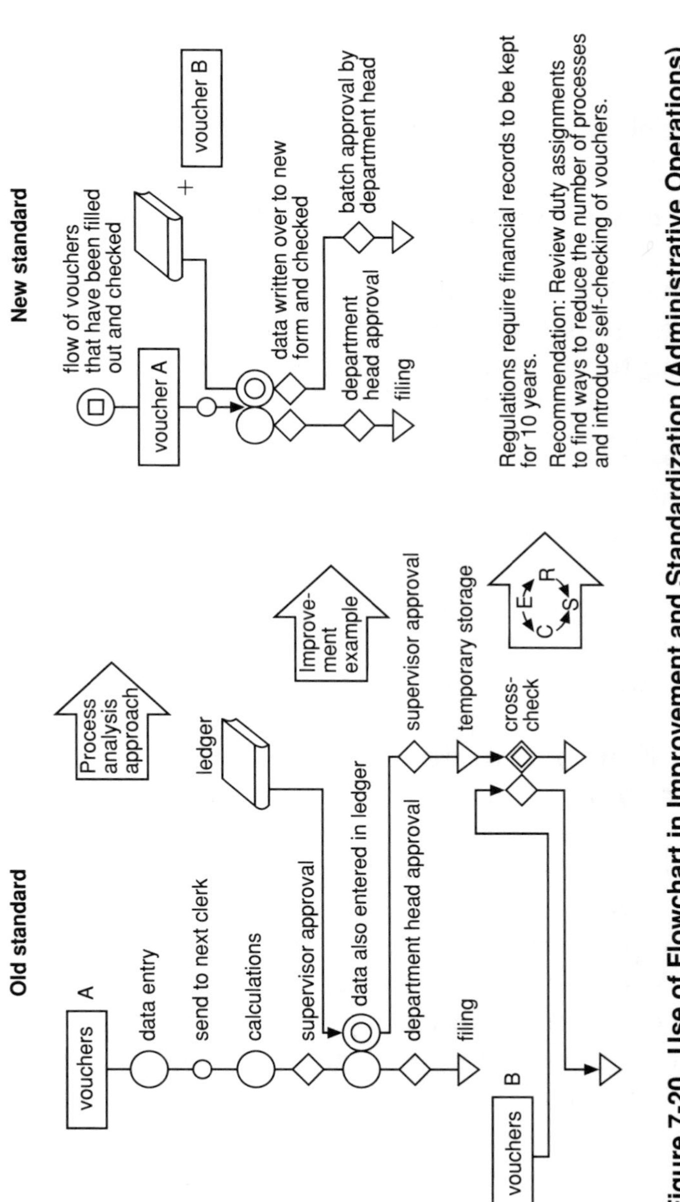

Figure 7-20. Use of Flowchart in Improvement and Standardization (Administrative Operations)

production management has come to rely more heavily than on computer-based information management systems, such as the one in Figure 7-21.

Within this type of system, standard times are used for the objectives listed in Figure 7-22. Ordinary methods used for process control can serve these objectives, although a slightly different method, known as PERT, better serves project planning objectives.

Applying PERT in Furniture-Moving Operations

This case study involves a department of about 30 employees. The workers applied PERT (program evaluation and review technique) to plan a project that involved moving furniture and other items into a new structure. The group began thinking of ways to standardize their moving operations to make them go smoothly.

1. Requirements
   a. Everyone will be involved in the move, but the only time available is Friday from 1:00 to 3:30 P.M. and on the weekend.
   b. Employees at greatest risk of injury when moving heavy items will be assigned only lighter items to move.
   c. Most of the items are documents. If these are gathered into boxes with destination labels, a parcel service can deliver them to the new building on Monday. All desks should be locked and taped to prevent drawers from opening. Glass items and other breakables should be moved to the new building on the same day.
2. Data to be gathered before moving day
   a. Using a work-factor system to analyze container moving operations, the group determined that the time for loading books from shelves into boxes was two minutes per person per box.
   b. Documents can be stacked into containers as high as 120 cm. Higher containers would reduce the number needed but would make them too heavy and might result in excessive fatigue or injury.
   c. The group observed container-packing work done by other departments. They found that these departments checked the documents before packing them into containers. They estimated that this inspection step would add one employee-hour per container, so they decided against it.
   d. They estimated the total volume of documents to get an idea of the total labor-hours needed.

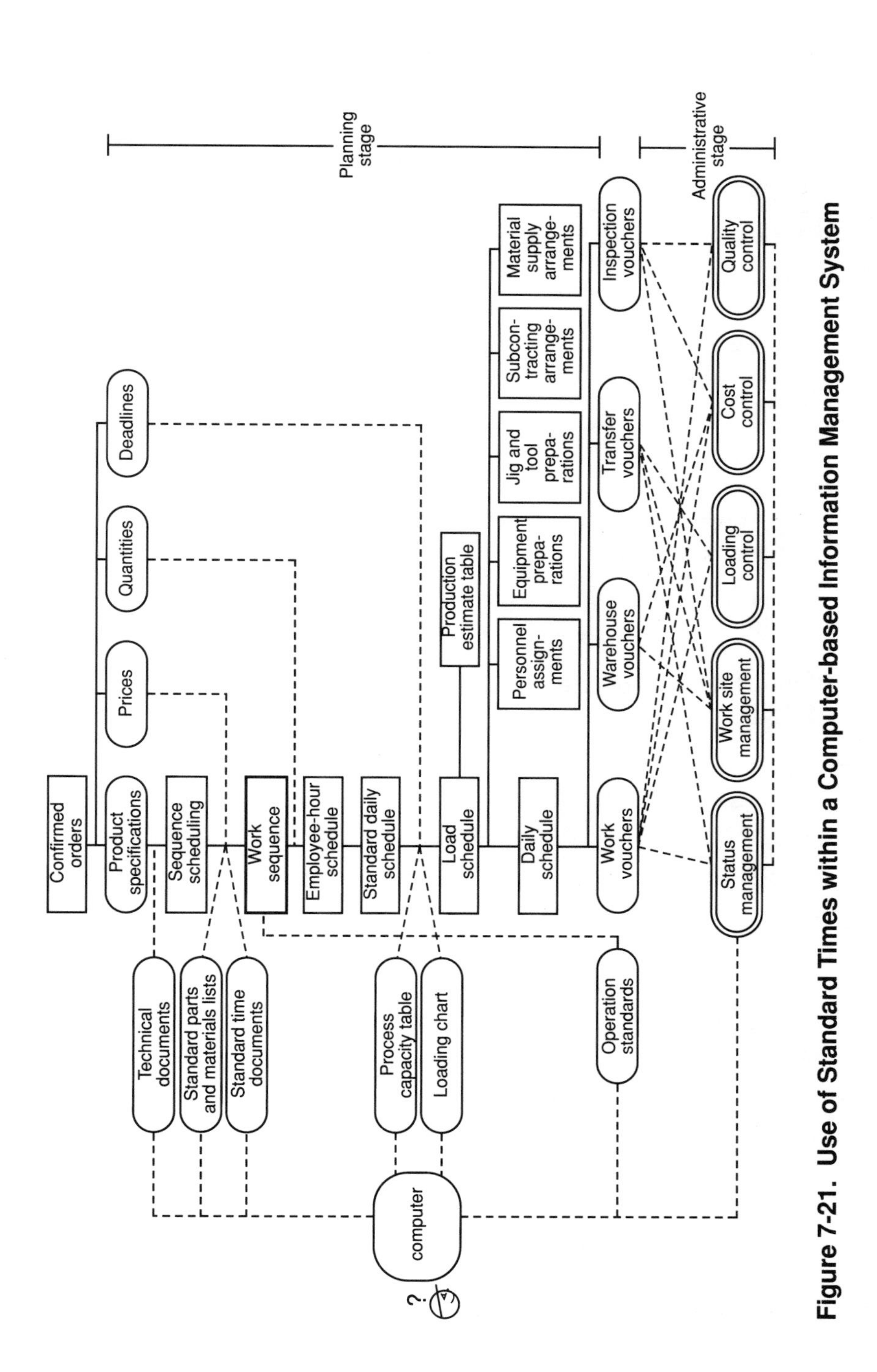

Figure 7-21. Use of Standard Times within a Computer-based Information Management System

Objective	Description
1. Estimation of processing costs	For estimating the processing and assembly employee-hours that form the basis for calculating production costs
2. Operation improvements	Provides a basis for comparing the levels of different manufacturing methods
3. Production schedule proposals	Used to determine the staffing and equipment requirements for production
4. Production management	Used in management of daily schedule, load, and capacity utilization
5. Subcontract management	Used to determine order prices, estimating the subcontract work ratio, and controlling subcontracted volume
6. Cost control	Basis for establishing per-unit labor costs, profit-loss calculations, and design revisions
7. Production efficiency measurement	Establishment of goal-oriented management to eliminate waste and policy formulation based on comparisons with other companies and departments
8. Salary bonuses, etc.	Reference for contract unit price, technical skill level evaluation, and evaluation for promotion
9. Delivery estimates and improvements	A basis for estimating the manufacturing schedule

Figure 7-22. Objectives for Standard Times

3. Preparations
   a. They estimated the standard time per box as (2 minutes × a rating of 200% × [1 + 25%] =) 5 minutes. The 200% compensates for the workers' age and slowness and the 25% for idle time. They then added one minute per box to the standard time as a margin for the learning curve.
   b. They made a detailed list of the items to be moved and divided the document containers among six work teams, as shown in Figure 7-23.
   c. They used an estimate of the total volume of documents to calculate the total labor required based on a lenient standard time, then they divided the job positions into groups (see square boxes in Figure 7-24).
   d. They made a PERT diagram (see Figure 7-25) based on the data in Figure 7-24. Two days before moving, the work team leaders met to go over the work sequence and the routes. For team leaders they chose capable workers who would actually be involved in the move.
   e. They displayed their PERT diagram and made a safety chart to encourage safe lifting and movement.

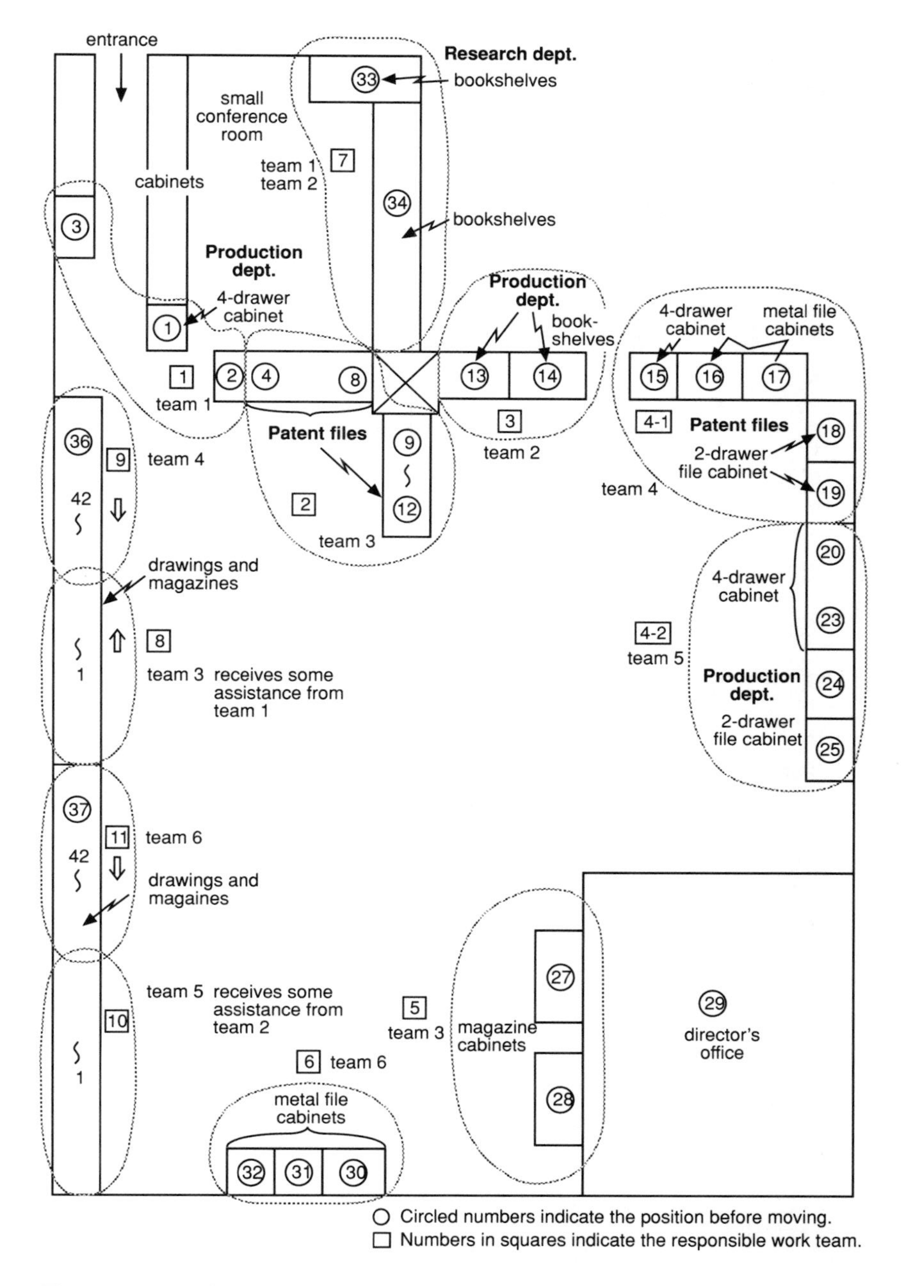

Figure 7-23. Standardization of Moving Operations (Division of Work in Removing Items from Office)

Position No.	Description	Length (cm)	No. of containers	Operation/ time
①	(Prod.) 4-drawer cabinet	80	2	1 21 min.
②	"	80		
③—1	Catalogs	60		
③—2	Microfilm	120	1	
④	(Spec.) 4-drawer cabinet	80	5	2 30 min.
⑤	"	"		
⑥	"	"		
⑦	"	"		
⑧	"	"		
⑨	"	"		
⑩	"	"		
⑪	"	"		
⑫	"	"		
⑬	(Prod.) bookshelves	650	11	3 66 min.
⑭	"	650		
⑮	4-drawer cabinet	1590	14	4-1 84 min.
⑯	Metal filing cabinet			
⑰	"			
⑱	2-drawer cabinet			
⑲	"			
⑳	(Dev.) 4-drawer cabinet	1590	14	4-2 84 min.
㉑	"			
㉒	"			
㉓	"			
㉔	(Prod.) 2-drawer cabinet			
㉕	"			
㉗	Magazine cabinet	450	9	5 54 min.
㉘	"	600		
㉙	Books	—	1	None
㉚	Glass cupboard	260	5	6 30 min.
㉛	Patent files	170		
㉜	"	170		
㉝	(Res.) bookshelves, small conference room	540	5	7 156 min.
㉞	Magazine shelves, small conference room	2160	20	
㉟	Desktop items, small conference room	80	1	
㊱ 1~21	(Eng.) Drawings — Entrance side	3549	15	8 90 min.
㊱ 42~22	" — Entrance side		15	9 90 min.
㊲ 1~21	" — Window side	3759	16	10 96 min.
㊲ 42~22	" — Window side		16	11 96 min.
—	**Total**	**17,108**	**150**	**897 min.**

Figure 7-24. Standardization of Moving Operations (Analysis of Container Packing Operations)

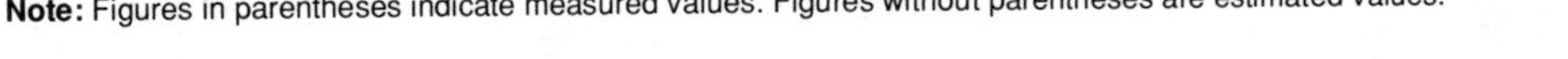

Work Team	Leader	Members
All teams	SN	All members
Hand truck team		SO IM
Carton loading team for No. 3 conference room	FJ	TK
Work team 1 Production files	SA	OG YM
Work team 2 Production files	"	FJ SK
Work team 3 Engineering files	YN	OT AK
Work team 4 Development files	NZ	ON TG
Work team 5 Development, Production files	"	NI MM
Work team 6 Measurement and loading	KR	KT SI
Finish up, progress check, and telephone report	SN	Staff
Desktop items	Staff	Staff

Note: Figures in parentheses indicate measured values. Figures without parentheses are estimated values.

Figure 7-25. Standardization of Moving Operations (PERT-based Time Schedule)

4. Sequence of operations on moving day
   a. The overall leader (SN) explained the safety points to everyone, then the employees split into work teams and reviewed the PERT schedule in detail with their team leaders.
   b. They moved the desks, established a path, and gathered the telephones in one place.
   c. They finished step 4b at 1:40 (20 minutes ahead of schedule), so they took a 10-minute break. The leaders reviewed the safety points, then began packing books into boxes.
   d. Some workers were fatigued after about 35 minutes, so they stopped and discussed the sequence of operations for about 15 minutes.
   e. They took another break after 40 minutes. People had been working at different paces, so some employees switched jobs for a while to even out the flow of work. Later, they took another 10-minute break.
   f. The work became more specialized as time went on. Most of the work was completed by about 3:40, when everyone took another break.
   g. Some people offered to pitch in with the remaining minor tasks, and were allowed to do so independently.
5. Bottlenecks
   a. One work team was reorganized because one member was sick that day.
   b. Some unforeseen remodeling work was being done, which interrupted one team's work, but this work was covered later.
   c. Staff employees assisted with supply shortages and other unforeseen problems to help the work go smoothly.
6. Evaluation of results
   a. Everyone was aware of the attention paid to the way each person did his or her work, and a spirit of competition grew among the work teams. This same visibility made it easier to discover inconsistency, irrationality, and waste in the work procedures; therefore everyone was keen on making improvements.
   b. One employee took time measurements and followed up on the PERT schedule to confirm the timing. The measurement and follow-up results were relayed back to the work teams, which helped them do the job confidently.
   c. The work team leaders showed strong leadership abilities. Knowing the team's work was based on reliable data, the leaders were confident in their work, especially in assigning jobs to team members.

d. All employees did their fair share of work and felt they had done it quickly and without overexertion. This came out during a follow-up discussion after the move, where everyone reflected on their achievement over refreshments.
e. At the new office, unpacking boxes and stacking books onto bookshelves went smoothly even without a PERT diagram. All the office employees got back to their regular work by noon on the Monday after the move. People in other departments were surprised at the speed and smoothness of the move.

Although this case study does not concern production activities, you can still learn a lot from it about establishing and using standard operations. The main points can be summed up as follows:

1. A detailed study of the job in advance makes it easier to keep execution on schedule. Methods such as PERT allow you to foresee the critical points (bottlenecks) and use your resources to avoid those problems.
2. Basing the work on theoretical standard times makes it easier to recognize inconsistency, irrationality, and waste.
3. Doing the work in small groups enables you to enjoy the camaraderie of planning and improving work procedures together. In this case, the work team members were not given detailed operation sequences at first, but they received detailed instruction from group leaders who understood the correct procedures.

The use of standard times worked well both directly and indirectly in the above case study. I chose this case study because everyone can understand moving operations. You can apply these methods to your own operations at other types of work sites.

AUTOMATING AND UPGRADING STANDARDIZATION

Referring back to the list of directions for standardization shown in Figure 3-4, I would add one more direction: pursue easier, quicker, and more reliable operations. Generally, this pursuit follows an approach that goes from people to methods to jigs and tools, and from the use of measurement tools and gauges to improved mechanization to materials. Several examples help clarify this point.

Direction for Upgrading Standardization

This example concerns an assembly line for electrical outlets (plug receptacles). Figure 7-26 shows the layout of the assembly work tables and other equipment. You can observe the following flow of work in these assembly operations.

Operator A walks from the work table to the wire bin, a distance of about three meters. The wires are precut and laid out in a box. The assembly operation requires about 20 wires, sometimes more when rework is needed afterward, so Operator A picks up a few more than 20 wires and walks the three meters back to the work table.

To make the next step easier, she lays out 20 wires on the work table, inspecting each wire as she counts and arranges them. Usually, at least one or two wires need rework (such as restripping); in such cases she puts those wires in the rework pile, which is also about three meters away from the work table.

Usually, even after sorting out the wires that need rework, she still has more than 20 wires left, so she returns the extras to the wire bin three meters away (she would rather keep them at the work table but cannot, due to lack of work space). To make her next step easier, she carries two small parts boxes with her directly from the wire bin to the bins of terminal connectors and outlet boxes (which are three meters away from the wire bin). She places one batch each of terminal connectors and outlet boxes into her parts boxes and carries them back to the work table. The distance from the outlet box pile to the work table is about five meters.

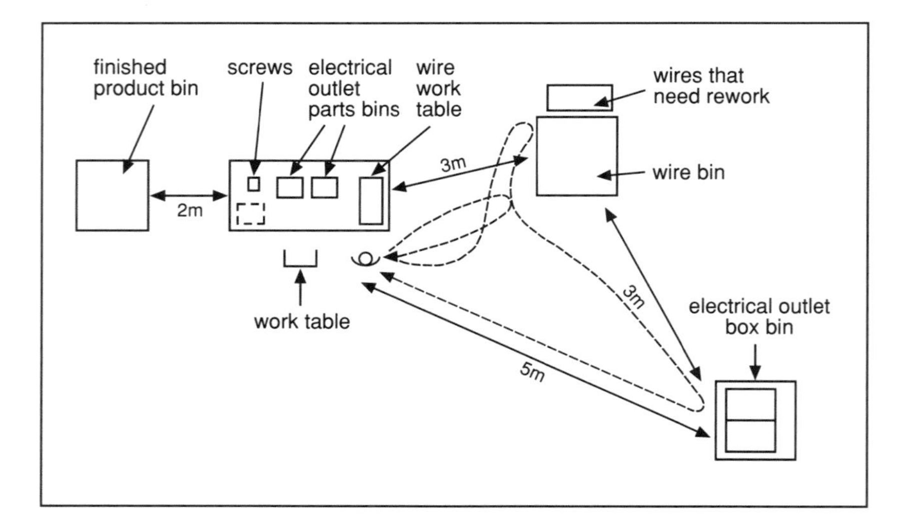

Figure 7-26. Layout of Electrical Outlet Assembly Operations

Seated at the work table, she uses a screwdriver to attach wires to the two terminal connectors, then picks up an outlet box, some screws, and other parts and assembles them. She places the assembled outlet at the left front corner of the table (inside the broken-line square in Figure 7-26). When she finishes 20 outlets, she gets up and moves them to the finished product bin (about two meters away), then walks back past the work table to go to the wire bin for another batch of wires. This completes one cycle of assembly operations.

Clearly, these assembly operations contain a lot of waste. If you look at the current conditions from the perspective of process analysis, you find that conveyance is the first area to improve. Figure 7-27 shows the results of an analysis of current conditions for these assembly operations. Figure 7-28 shows the corresponding improvement results and Figure 7-29 outlines general improvement targets.

The main improvement points include

1. The wires and other electrical outlet parts move closer to the operator's work table.
2. The wire remains on a reel to make it more easily accessible.
3. A foot-operated wire cutter frees the operator's hands for other work.
4. The operator can pick up all needed parts and perform the assembly work while seated at the work table.
5. The finished product bin has been moved next to the work table.

Most of these changes are operation method improvements, but they also include some equipment improvements, such as the introduction of the wire cutter and certain jigs and tools.

At the next stage, an automatic parts feeder could be installed to supply the terminal connectors and an auto-fastening device could be used to attach the connectors to the wires. Installation of a mechanism that transfers the assembled wire and connector into the case would be well on the way to fully automated assembly. Naturally, full automation is not worth the heavy equipment investment unless the production output is quite large. Operations with more modest output levels have to make do with additional equipment (i.e., mechanization) instead of automation.

The next level of improvements deals with materials. To understand what this means, we reexamine the electrical outlet assembly operations, using VE (value engineering) methods. Figure 7-30 describes the operations' objectives from the "Step" column in Figure 7-27 using the VE method of functional family tree development. This kind of diagram can serve as a helpful reference when devising improvement ideas for materials.

Step	Work	Transport	Hold	Inspect	Distance (m)	Time (min.)	4W1H					Comments
							What?	Where?	When?	Who?	How?	
1. Get wires	○	•	▽	□	3	0.04		√				Can wire bin be placed closer to work table?
2. Pick up at least 20 wires	●	∘	▽	□		0.20	√					What if she picks up exactly 20 wires?
3. Carry wires to worktable	○	•	▽	□	3	0.04		√				What if the wires were cut at this process?
4. Inspect and arrange wires	●	∘	▽	■		0.02	√				√	
5. Return extra wires	○	•	▽	□	3	0.04		√				
6. Go to terminal connector bin	○	•	▽	□	2	0.03		√			√	Can a cart be used to carry sets of parts and wires?
7. Place connectors and outlet boxes in parts boxes	●	∘	▽	□		0.20	√					
8. Return to worktable	○	•	▽	□	5	0.06		√			√	
9. Assemble 20 outlets	●	∘	▽	□		10.00	√					
10. Carry outlets to finished product bin	○	•	▽	□	2	0.03		√				Can this be moved closer to the work table?
11. Return to worktable	○	•	▽	□	2	0.03		√				

Figure 7-27. Analysis of Current Conditions in Electrical Outlet Assembly Operations

Step	Work	Transport	Hold	Inspect	Distance (m)	Time (min.)	4W1H					Comments
							What?	Where?	When?	Who?	How?	
1. Cut wires	●	○	▽	■		0.02	√					Use a foot-operated wire cutter
2. Pick up terminal connectors and assemble electrical outlet	●	○	▽	□		0.30	√					
3. Place assembled electrical outlet in finished product bin	●	○	▽	□		0.01	√					

Note: The wire cutter blade must be checked every day. When wire on the reel runs low, an alarm lamp signals the need for a refill.

Figure 7-28. New Standards for Electrical Outlet Assembly Operations (After Improvement)

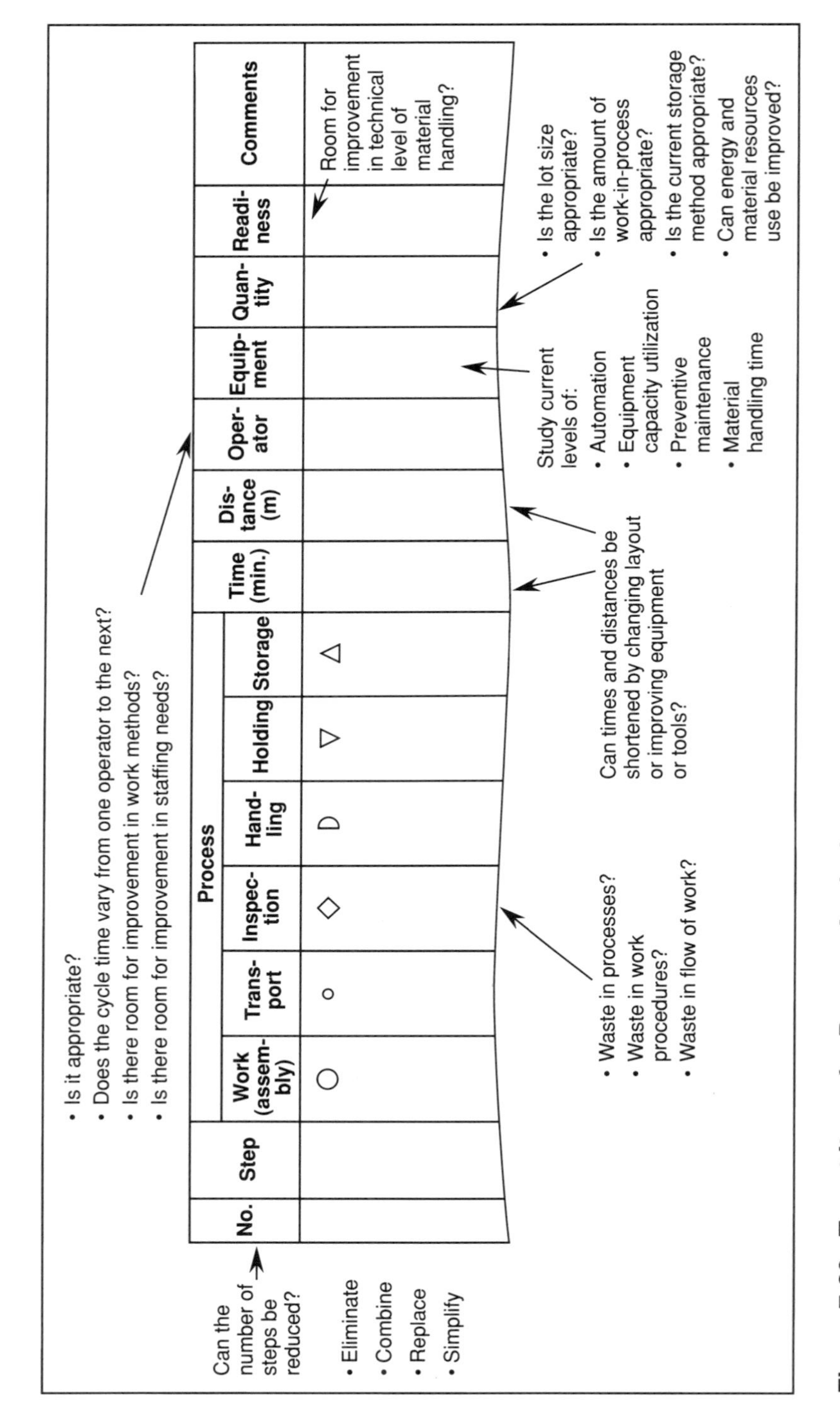

Figure 7-29. Target Items for Process Analysis

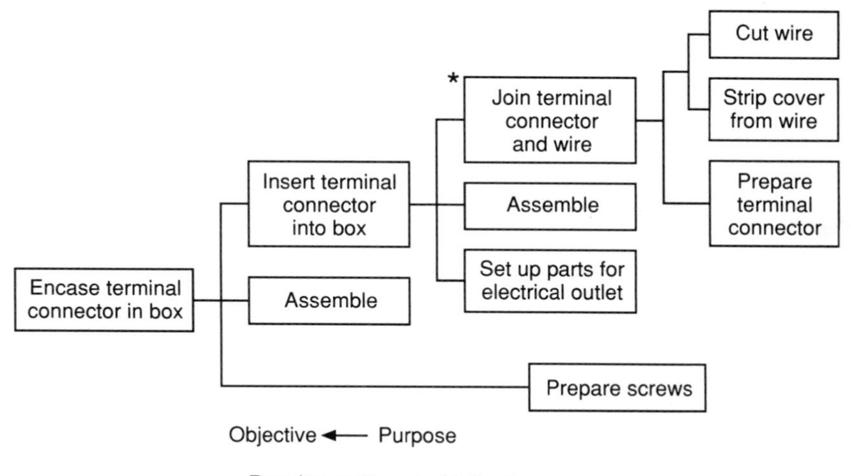

Figure 7-30. Functional Family Tree for Electrical Outlet Assembly Operations

Let's try to come up with improvement ideas for the "subject" and "verb" functions in this functional family tree. Figure 7-31 lists some ideas regarding the function described as "join terminal connector and wire" (see asterisk in Figure 7-30). We can think of various ways of joining these two parts, such as caulking or welding. These alternatives may offer an improvement over the current method.

As for "encase terminal connector in box" (see box at far left of Figure 7-30), ideas might be as simple as a plastic or rubber clip, or as sophisticated as automating the fastening of terminal connectors instead of using screws. Always consider the objective; in this case, "Why must the terminal connector be encased in the box?"

Although not shown in Figures 7-30 or 7-31, you might think ahead to the next process and come up with a more extensive improvement, such as "join circuit A and circuit B" using a single baseboard on which to connect the two circuits. This eliminates the need for a plug and outlet, which is a significant improvement. It eliminates the labor-hours spent inspecting and reworking defective outlet parts, which in this case is a major achievement in labor-hour reduction.

As we take your improvement ideas farther up the functional family tree, we eventually reach the design stage. Ideas for standardizing equipment and

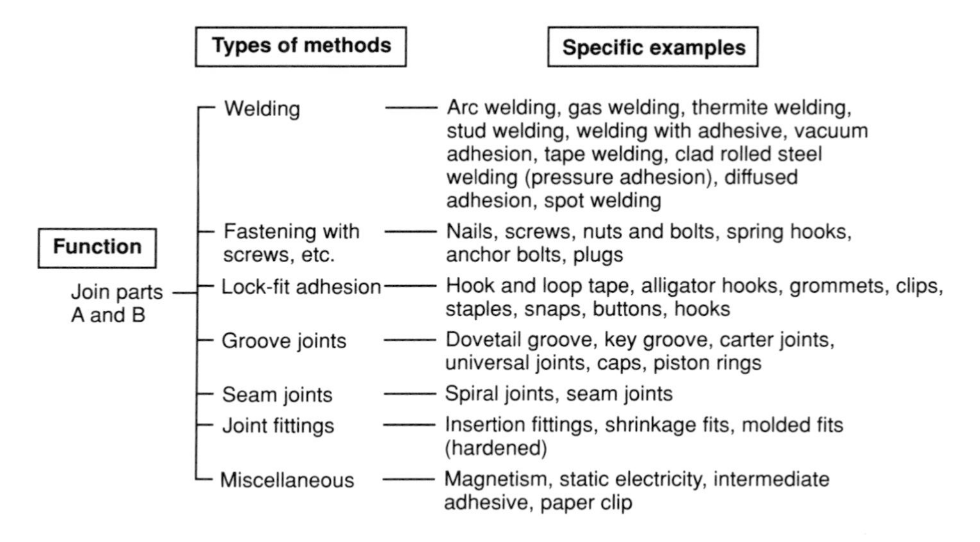

Figure 7-31. Alternative Ways to Join Two Parts

improving materials can be applied at production work sites as well as in the design of components and products.

Simplify Manual Operations before Automating

In addition to the more sophisticated operation improvement techniques described in earlier sections, I would also add a few words regarding the mechanization of more familiar operation procedures. Most cases of mechanization fit one of two categories:

1. *Mechanization within the framework of current operations.* This usually requires the installation of complicated mechanisms on the currently used equipment. Such improvements usually cost a lot and break down (although they appear technologically sophisticated).
2. *Mechanization after thorough improvement and simplification of current operations.* This is a simpler approach to mechanization. It costs less and breaks down less frequently. Such simplification requires repeated analyses and improvements.

Both categories have their relative advantages and disadvantages, but I recommend the second approach to mechanization whenever possible.

At a certain factory, one of the lathes produced a lot of smoke whenever it was used for cutting, so the managers quickly responded by installing smoke exhaust equipment; however, this new equipment kept breaking down. As a further improvement, they posted some maintenance instructions on the equipment and installed a breakdown detector.

The root cause for the smoke was the scattering of hot filings into the cutting oil. Therefore, a simpler and more effective improvement would have been to cool down the cutting area (such as by using a blower that supplies −10° C air to the cutting surface) and to replace the current cutting oil with burn-proof oil. These simpler countermeasures would have eliminated the smoke problem, avoiding the need for expensive and unreliable smoke exhaust equipment.

Moving from the problem of smoke emission to smoke exhaust equipment is a quick "improvement," but not a very smart one. Unfortunately, the mechanization and automation efforts going on at factories today include many such ill-considered shortcuts.

I stress again that the correct approach to mechanization and automation begins with a firm grasp of the principles involved and then works to simplify current manual operations before thinking of ways to mechanize or automate them. Avoid mechanizing current operations just the way they are; it can be an expensive "shortcut."

8
Standardization for Production Management

A CHANGE IN PRODUCTION MANAGEMENT OBJECTIVES

Production management manages people, materials, equipment, and other resources needed for production activities, including information. Production management obtains the fullest and most efficient use of production resources to help meet the company's objectives.

From the perspective of creating or adding value, corporate objectives improve the rate of earnings on total capital (T), which breaks down as follows:

$$T = \frac{\text{Net profit}}{\text{Total capital}} = \frac{\text{Net profit}}{\text{Total sales}} \times \frac{\text{Total sales}}{\text{Total capital}}$$

= Earnings rate on total sales
= Sales improvement (improved earnings rate on total sales)
× Turnover improvement (improved total capital turnover)

The sales approach mostly uses cost-cutting measures. The first half of Chapter 5 explains several of these and will not be repeated here. In a nutshell, these measures help the company make more efficient use of its people, materials, and equipment to make nondefective products at lower cost.

The turnover approach uses those measures designed to shorten the production period. These measures also deal with people, materials, and equipment, but the objective here is to find the most efficient combination of these three elements to make production activities as waste-free as possible and to speed up production turnover. All production activities depend on maintaining good control of the *what, when, where,* and *how many* variables of production, with a view toward transforming materials into products in the shortest time possible. Production management is about maintaining such control. It helps to think of production management functions as P-D-C management functions, such as those listed in Figure 8-1.

Most importantly production management helps smooth the flow of activities that begins with receiving customer orders and ends with shipping products to customers. Production management sometimes extends even farther upstream, to shorten the periods needed for the design and commercial development of new products. Recently, shortening the lead time between receiving orders and shipping products has been a key marketing strategy for companies establishing just-in-time (JIT) or computer-integrated manufacturing (CIM) production management systems. (We should note that these strategies go beyond the production system and include distribution systems as well.)

Why Emphasize Production Turnover?

Why do management strategists insist on shortening lead time between receiving orders and shipping products? Consider, for example, how different kinds of sandwich shops would look at the sales and turnover approaches described earlier.

Sales approach. A restaurant at a luxury hotel would probably take this approach. This kind of business charges high prices and does not sell very much but has a large profit margin on each sandwich plate sold. Profit is based on high prices and low sales volume.

Turnover approach. A hot dog stand at a train station would probably take the turnover approach. Such a business has a low profit margin on each hot dog sold, but is not the kind of comfortable place where customers linger after eating. Consequently, customer turnover is high. Profit is based on low prices and high sales volume.

The restaurant business tends to be polarized between low-volume luxury and high-volume no-frills businesses. The high-volume no-frills places pay roughly the same prices as the low-volume establishments for ingredients, equipment, and labor; to make a profit on a low-price menu they must have high customer turnover. High-volume businesses must buy ingredients more frequently when sales are brisk; when sales slow down, they may be stuck with large amounts of unsold perishable goods. Sometimes this kind of business fails

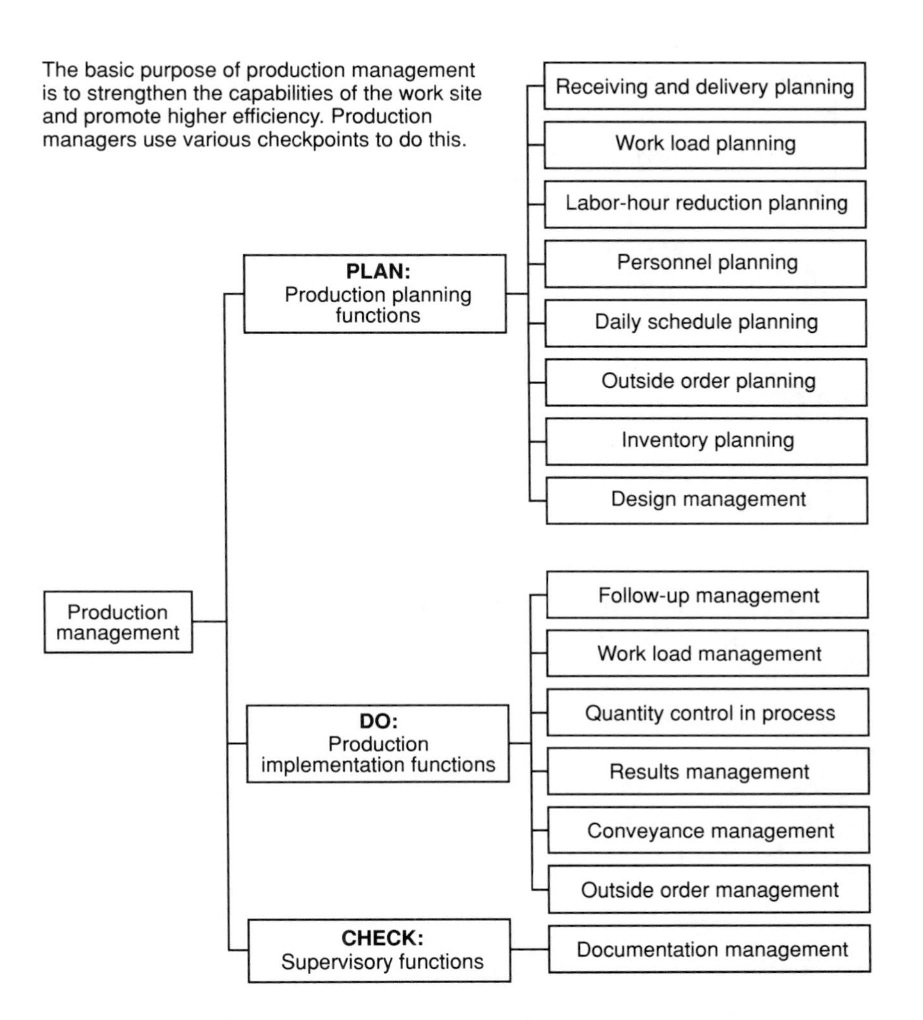

Figure 8-1. PDC Functions and Management Themes for Production Management

to sell enough sandwiches to cover the cost of fresh ingredients. Accordingly, in using the turnover approach, you must recognize the importance of controlling capital turnover and seek to accelerate the turnover to enhance profitability.

The management policies of each approach can be summarized as follows:

- *Sales approach:* Develop high-grade products to serve luxury-minded customers. Seek out a small but steady clientele. Maximize the profit margin on materials by reducing labor costs, procuring higher-quality materials at similar or lower costs, and reducing other costs. This requires a high level of management skill.

- *Turnover approach:* Keep product prices competitive with those of local rivals, reduce inventory and shorten production lead time to more readily meet customer needs. Improve the efficiency of equipment changeover and setup to make more efficient use of available space.

With the rapid waves of change washing across today's business world, most companies are emphasizing the turnover approach. The sales approach is still the better approach for businesses that make or sell art objects, luxury products, and designer-brand products, but even these companies are beginning to pay more attention to turnover-approach policies.

In the sandwich shop example of the turnover approach, the central management policy is to accelerate capital turnover to improve the rate of profit on total capital. In most companies, managers also emphasize the following reasons at various times:

1. By having a rapid capital turnover, even with a low profit rate (a stable rate of profit on total sales), the company can obtain a high profit on total capital when turnover increases sales. This was explained in the sandwich shop example above.
2. The company's inventory includes materials kept at every stage from warehouse receiving to product shipping (this inventory is called inventory assets). Interest must be paid on the capital invested in inventory assets, and the best way to reduce the interest burden is to reduce the amount of inventory. This prevents the accumulation of dead inventory, such as materials rendered obsolete by product model changes, and makes the company more responsive to changes in product life cycles.
3. Shortening the delivery period provides an intangible service to customers and adds value to products.
4. A shorter production lead time enables the company to bring new products to market faster and thereby gain an advantage over the competition.

The first two reasons are easy enough to understand, but the third and fourth may need more explanation.

The third reason, shorter delivery periods, is actually a significant service for customers. From the customer's viewpoint, it is usually best to have products shipped as soon as they are ordered.

For example, a customer orders materials, parts, or equipment for developing new products or expanding its production facilities. Each day eliminated from the delivery period gives the customer another day's head start in bringing its new products to market or its new facilities on-line. The assurance of speedy parts delivery also gives the customer more time to study market trends before committing to produce a certain product, which can be a valuable service.

Such an advantage is especially important in today's fast-changing times. Market conditions can change dramatically over the course of a few short months. This poses a big risk to manufacturers who must invest large sums in materials and parts for each product they turn out.

As for the fourth reason, shorter production lead time, its benefits can significantly improve a company's internal competitiveness. Shorter production lead times enable companies to increase the amount of feedback received during production.

Sports enthusiasts enjoy instant appreciation of the results of their efforts. Even in golf, the golfer can see immediately whether a new swing technique works. Accordingly, he or she can repeatedly experiment with various techniques to find the most effective one. Instant feedback enables anyone to become better quickly as they frequently practice a new skill. On the other hand, when a person practices infrequently, that person must spend more time returning to the starting point and reviewing what was practiced earlier before he or she can learn more. This also slows the rate of feedback on new methods and therefore slows progress.

The more frequently you can repeat the cycle of practicing (learning a new method) and testing (evaluating the effectiveness of the new method), the faster you improve and the greater your competitive strength becomes.

Similarly, any company that has shortened its production lead time has also speeded up its rate of improvement and its rate of growth in competitiveness. The company can more quickly find out which products are good and can resolve problems in inferior products more quickly. In a nutshell, shorter production lead times make stronger companies.

You can see, then, how important the turnover approach can be. Look at any company that is busy using computers or other advanced methods for shortening production lead time and accelerating the flow of feedback information, and you will see a company motivated by the fourth reason for pursuing the turnover approach.

Many companies, in fact, are introducing computer-based production management methods in an attempt to keep in closer touch with customer needs and move closer to the forefront in opening new markets. These companies, in particular, should pursue all four reasons for the turnover approach as they seek to build stronger production systems.

Small-Lot Production That Can Adapt to Change

Traditionally, wide-variety, small-lot production employs conveyor-based production using the one-piece flow system, while small-variety, large-lot production uses more of a job-shop layout, with machines grouped by type, not by process. Production activities have changed so much in recent years that some younger workers have never heard of the job-shop system. Ideas that were once revolutionary, such as quick changeover and flow-oriented production management, are now the norm in some industries.

Figure 8-2 shows examples of measures taken to establish small-lot, leveled production. The bottom part of the figure describes the conditions that exist after establishing small-lot, quick-changeover production. These data are based on a production lead-time reduction to one-fourth of the previous time. The company achieved this reduction by following the turnover approach and the four reasons described earlier.

Note that the total monthly output — 30,000 units — stays the same because the company compensates for reductions in production output of one product model (item A, for example) by increasing the output of another model (such as item C). Therefore the total monthly output remains stable. The result is adaptability to change within the framework of a leveled production system.

Looking at it from a different perspective, this system can entirely replace product model A (of which 12,000 per month were produced via 30 lots of 400 units) with a different product model, although it would have to meet various conditions concerning materials supply lead time, standard times, and so on. Wide-variety, small-lot production sometimes requires such abrupt changes in production scheduling.

The key, however, is to maintain the same total output even while varying the output of individual product models. Factories can do this by establishing a wide-variety, small-lot, leveled production system.

Leveled Production Uncovers Abundant Latent Losses

This section looks at what leveled production means from the standpoint of the work site.

1. A leveled production system works with carefully determined small amounts of work-in-process and lot size for each process. As the amount

1. Production conditions
   - Production output: 1,000 units/day
   - Line enables mixed production of up to three different product models
   - Line operates on one 8-hour shift per day, 30 days/month

2. Previous production system

Model	Request from downstream processes	Day 1	Day 2	Day 3	Day 4	Day 5	Output/ month	Total production days	Average work-in-process
A	400 units/day	1,000			1,000		12,000	12	1,500
B	200		1,000				6,000	6	900
C	400			1,000		1,000	12,000	12	1,600
—	1,000	1,000	1,000	1,000	1,000	1,000	30,000	30	4,000

3. Small-lot quick-changeover production system (leveled production)

Model	Request from downstream processes	Day 1	Day 2	Day 3	Day 4	Day 5	Output/ month	Total production days	Average work-in-process
A	400	400	400	400	400	400	12,000	12	500
B	200	200	200	200	200	200	6,000	6	300
C	400	400	400	400	400	400	12,000	12	500
—	1,000	1,000	1,000	1,000	1,000	1,000	30,000	30	1,300

Advantages of new production system
- It can respond easily to variation in product output.
- It has a shorter lead time, less inventory, and makes better use of available space.
- It provides daily feedback of quality information.

Figure 8-2. Measures to Establish Small-lot, Leveled Production

of work-in-process is reduced, latent problems at the work site become more apparent.

2. The leveled production load makes it easier to schedule the use of labor, material, and equipment resources. Reorganization of personnel assignments is not required within the leveled production period.
3. The rapid turnover rate (frequent repetition) of production operations enables operators to learn and improve their skills more easily. It also enables preventive maintenance to be carried out more systematically.
4. New products are introduced in small lots also, which reduces their impact on the overall production system and subjects them to more frequent testing.
5. Leveled production systems are conducive to simple management methods (such as kanban for production management).
6. They are also conducive to the just-in-time approach whose goal is to produce only what is needed, only when needed, and only in the amount needed.

I would like to make particular note of the relation between leveled production and the uprooting and elimination of waste. Waste results when a company has periods of peak demand for labor, material, and equipment resources, followed by valleys when these demands are low (see Figure 8-3). Although it may be difficult to detect, the waste of resources during such sluggish periods can be great. Problems, although undetected, are still problems; they may not only hinder the company's future development but also drain the company's finances.

The main objectives of production management today are to eliminate waste in the use of labor, materials, and equipment resources; to find an efficient combination of these resources; and to shorten the lead time between receiving customer orders and shipping products. The path toward achieving these objectives includes taking the turnover approach and establishing small-lot, leveled production with minimal work-in-process.

This requires companies to make various improvements aimed at shortening the production lead time as well as to establish various standards, such as standard times, and to plan ahead for waste-free production scheduling. The list of production planning functions in Figure 8-1 should make a useful reference in this endeavor.

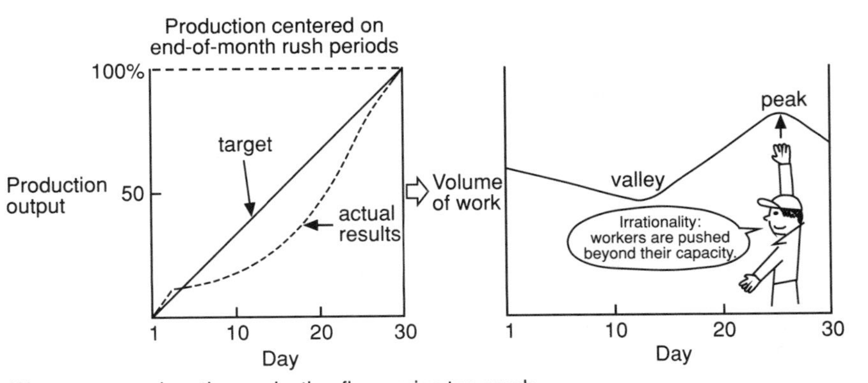

Waste occurs when the production flow varies too much.

Example: Labor, material, and equipment resources are overworked during peak periods and underused during slack periods.

Figure 8-3. Non-leveled Production Creates Irrationality and Waste

A FULL-FLEDGED CAMPAIGN TO SHORTEN PRODUCTION LEAD TIME

Figure 8-4 describes production lead time and its constituent elements. Improvement methods that shorten production lead time are targeted at the flow of information and make use of process analysis.

This approach begins by analyzing the "paper lead time," which includes all the information-based processes between the sales planning stage and the order reception stage. It goes on to analyze the manufacturing lead time, which includes all processes from the procurement (material purchasing) stage through manufacturing to product shipping.

Lead-Time Shortening Improvements for Information-Based Processes

Whenever "information" crops up as an improvement target, people tend to lunge toward computerization as the answer. Before exploring the possibilities of computerization, however, you should analyze the information-based processes, making waste-eradicating improvements, and standardize the improved processes. Only after you have done these things should you begin to consider how to divide the work between people and machines. This is a more rational approach to improving information-based processes.

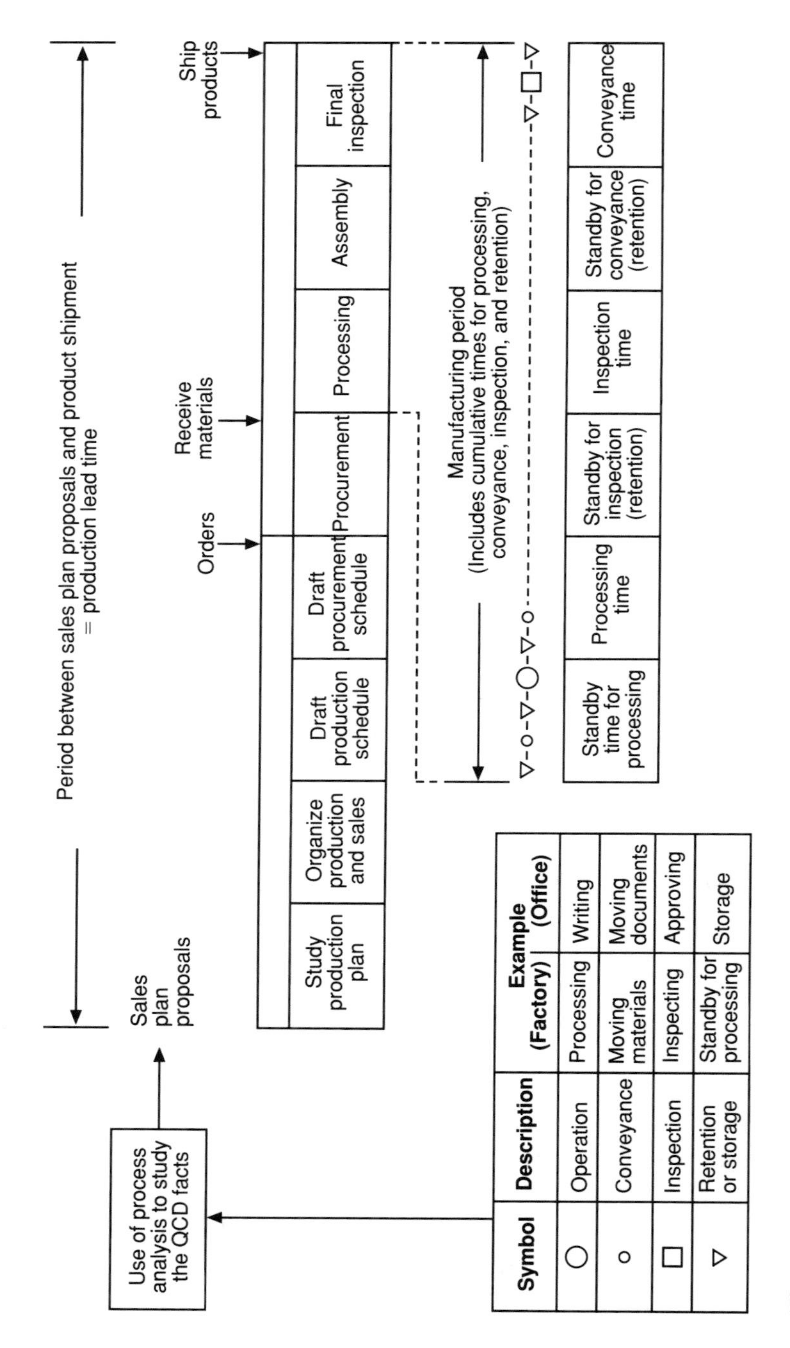

Figure 8-4. Production Lead Time and Its Components

Many methods can be used in this approach. Administrative process analysis lends itself well to analyzing information-based processes; we will look at how it can be used for production work sites as well as for offices.

This type of analysis uses some of the same process analysis features described earlier. The main difference is that instead of targeting the flow of materials or operations, it targets the flow of information. Figure 8-5 lists specific uses of information-based process analysis.

Improving Production Lead Time

To improve the production lead time, we must take a multifaceted approach that includes the following measures:

1. Using JIT approach to delivery of parts and materials
2. Improving links between processes
3. Synchronizing lines
4. Developing quick changeover

Steps	Waste elimination approach	Direction of countermeasures	
		Human work	Machine work (and potential machine work)
Operation	Rationalization via ECRS investigation and 5W1H analysis	Surveys, planning, investigation of order data and production schedules, improvements, responses to problems, etc.	Routine tasks such as receiving orders, planning production, compiling results statistics, and making photocopies
Evaluation	Evaluation of operations. Use of alarms and mistake-proofing devices.	Checks, inspections, and audits.	Alarm processes, mistake-proofing devices (parts delivery, warehouses, etc.)
Transport	Apply ECRS rationalization to transportation routes. Make routes wider and shorter.	Memos, instructions, dispatches, reports, meetings, etc.	Unify information format and mechanize data search functions.
Storage and standby	Reduce volume of documentation, use search index codes, etc.	Paperwork reduction campaign, implement 5S for offices	Voucher management, information storage, ledger management

Figure 8-5. Uses of Information-based Process Analysis

5. Improving equipment troubleshooting measures, preventive maintenance, and cycle time to eliminate bottlenecks in production processes
6. Standardizing and unifying parts
7. Managing storage sites (devices to prevent overproduction)
8. Standardizing operation procedures and skills improvement
9. Maintaining standard times
10. Improving responses to defects and rework (such as via poka-yoke devices)
11. Improving other measures (visual management, computerization of production management for more efficient processing of information, etc.)

You can shorten production lead time by making the flow of materials smoother. Storage points obstruct the flow of materials. This flow begins with input, involves processing and conveyance, and ends with output. Improvements in the flow of materials should focus on places where materials tend to accumulate and should identify and eliminate the root causes of such accumulation. For example, equipment breakdowns and other quality problems often cause poor processing and conveyance of materials when no problems exist at the input side. Also look at the upstream processes where materials tend to accumulate.

Chapter 5 examines specific flow-improvement measures that deal with materials, equipment, and operation methods. In addition to these measures, companies must also make full-fledged efforts to shorten production lead time.

CAUTION POINTS REGARDING COMPUTERIZATION

Today, computer technology has advanced to the point where companies naturally seek computer support for information processing to make production faster, easier, and more reliable. In particular, the production manager's job can be made much more efficient if she has access to only the necessary information, when it's needed, and in the amount needed. In addition, the changes that computerization brings to management work have spinoff effects that help raise efficiency in other departments as well.

Figure 8-6 shows an example of computerized production management operations in the process industry. This figure contains an information-processing flowchart that begins with the order reception stage and ends with the

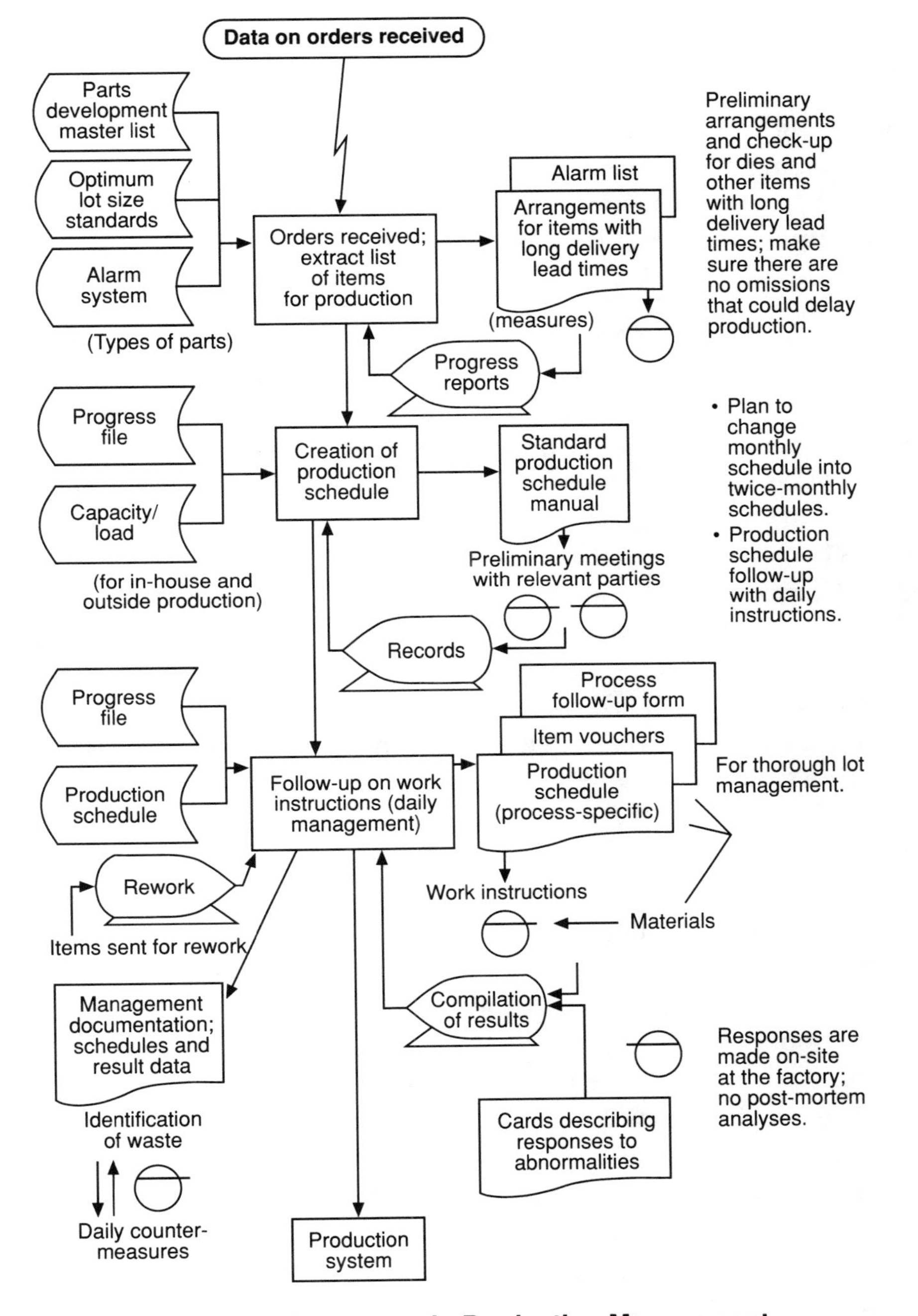

Figure 8-6. Use of Computers in Production Management (at a Process Industry Company)

production order stage. They sought to make use of computer applications that serve the purposes of

- unifying the data format.
- organizing and checking the alarm list to make sure there are no omissions.
- making instructions more reliable, based on the daily production schedule.
- providing swift compilation of results data to enable earlier detection of problems.

Figure 8-7 takes a closer look at daily production management. Most importantly, daily management must pay close attention to items that tend to cause delays and provide clear instructions as to when each unit of production equipment operates and what (and how many) items it handles. Daily management means checking daily how well these instructions are being carried out and taking immediate action to prevent today's problems from recurring.

The ST (standard time) is the foundation for these types of management activities. Comparing RT (result time or actual production time) values with ST values is a fundamental part of any kind of production management. In the example shown in the figure, you can see how work instructions (Plan) are written up as work results (Do), which are monitored daily (Check). This thorough approach to production management ties in directly with the planning and implementation of improvements.

Figure 8-8 shows how visual management methods, such as alarm lamps, computer monitors, kanban, and color coding, can support thorough daily production management to ensure a smoother production flow. These nuts-and-bolts methods take in minor problems and help fuel initiative for improvements and preventive action. You can use methods such as these to build a comprehensive production management system.

Remember that using computers is just one set of methods among many. The key is ensuring that the method you use is the most appropriate one for your management objectives.

Although the examples of computerized production management have been restricted to process-industry companies, this same management system can also be applied to other companies, such as those in the assembly industry. For assembly processes, this system emphasizes the management of parts supplied to the assembly line.

Various methods exist for managing the ordering and delivery of parts. In addition to the kanban system, there is the MRP (materials requirements planning) approach. Descriptions of MRP methods can be found in other

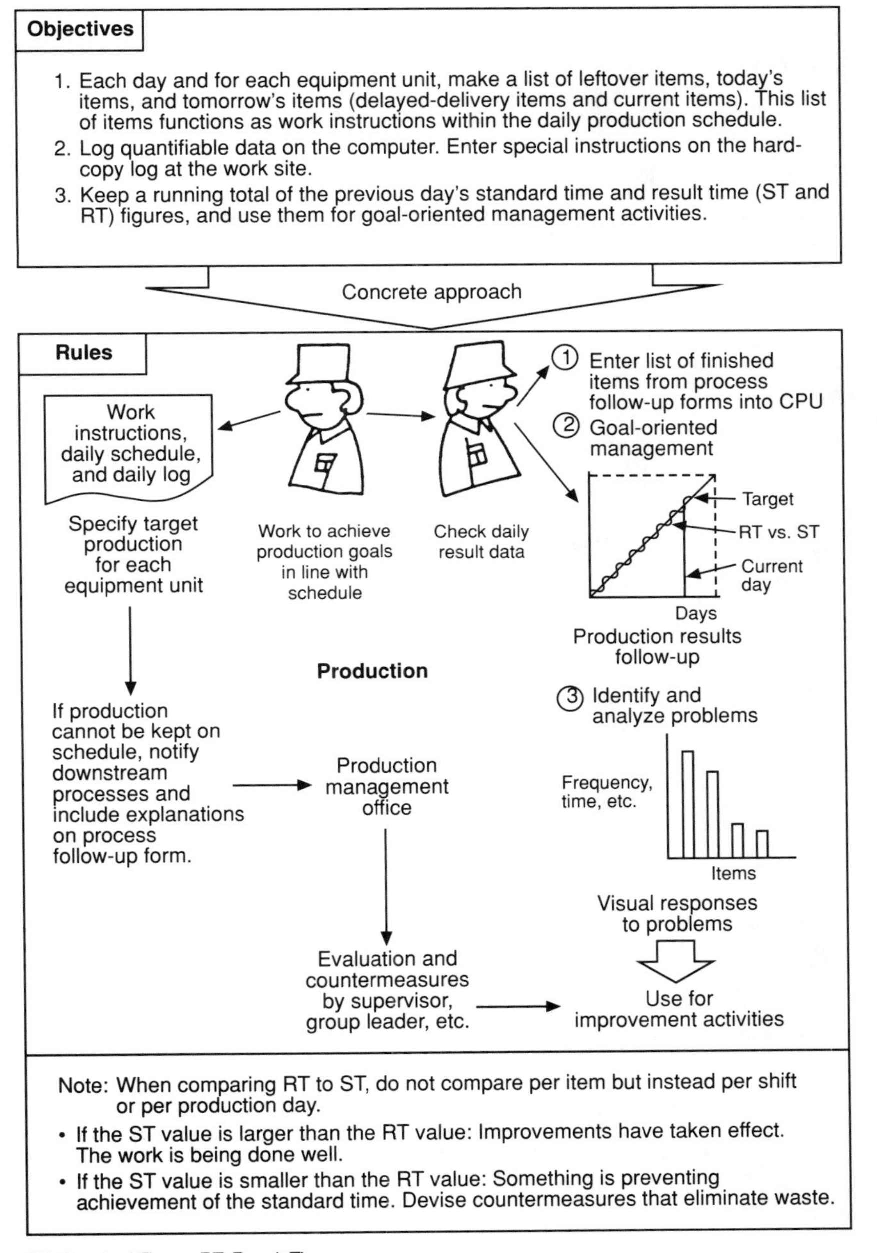

Figure 8-7. The Plan-Do-Check Cycle at Work in Daily Management

Key Points of Visual Management (kanban and display boards)
1. Visual displays reduce employee-hours required to communicate problems 2. Worksite displays lead to quick responses to worksite problems 3. Use everyone's input and wisdom. Make preventive measures the norm: do not gather data simply for post-mortem analyses.

Key Uses of Visual Displays

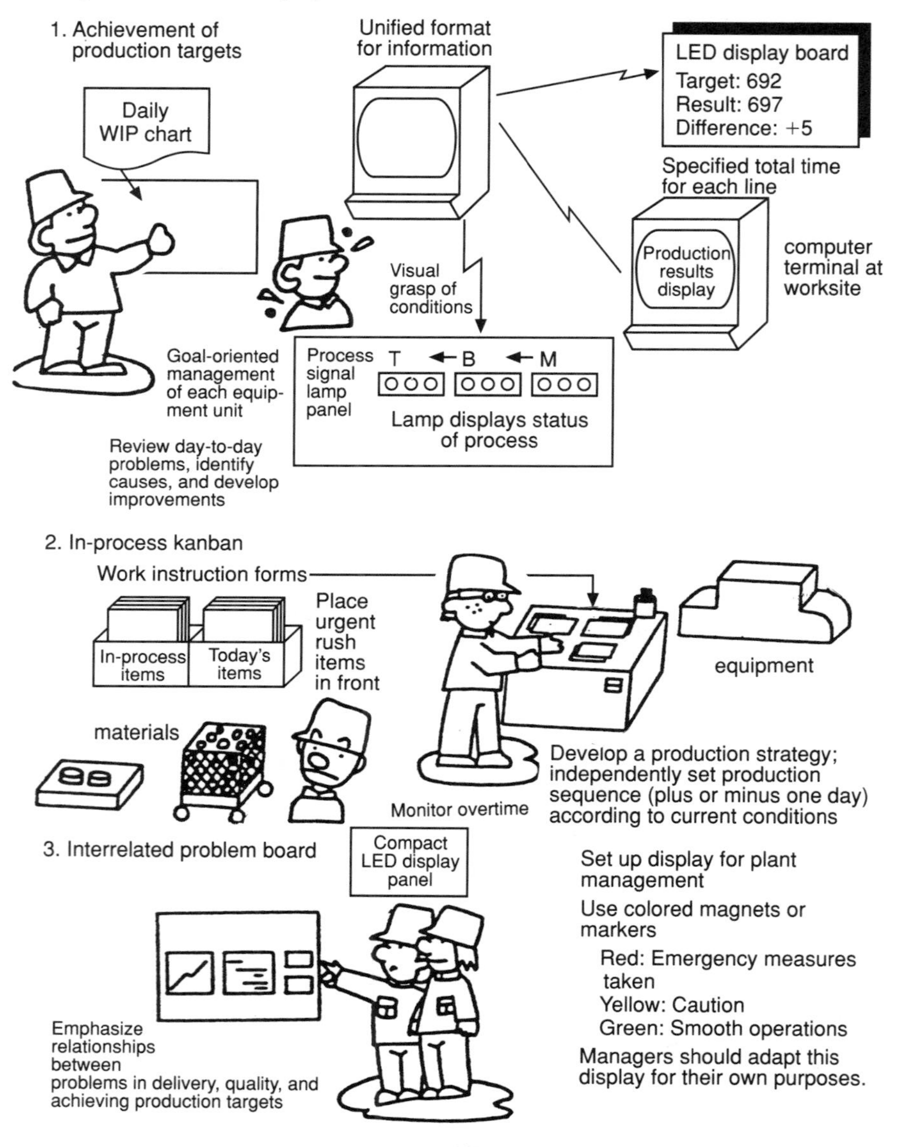

Figure 8-8. Visual Management Methods for Thorough Daily Management

books, so it will suffice to say here that this approach focuses on what, when, in what amount, and where parts need to be delivered. No matter which method you choose, you should have standardized methods for displaying and communicating information about production operations, equipment, and materials at the work site.

MANAGING TROUBLESOME PRODUCTION PROBLEMS

One of management's basic responsibilities is to respond quickly to problems. Instead of taking a passive stance toward such problems and assuming that no news is good news, production managers should inform workers that they want to hear any bad news, as soon as possible.

A good production manager tries to understand and solve tough problems quickly. Such troubleshooting is a major part of the job, along with predicting future problems. Managers must therefore focus their attention and energies on problem solving as well as resourceful planning.

At most companies, certain departments would like to see their colleagues in other departments do things differently. Figure 8-9 lists some of these common interdepartmental complaints. Some of these complaints may be painful for production managers to hear, but they nevertheless need to understand them if they truly intend to be good problem solvers.

I recommend the problem-solving system outlined in Figure 8-10. This approach sorts problems into three types, each with its own set of countermeasures. Under Type A, we establish an emergency response — rules for handling problems that require immediate "first-aid" countermeasures — and use provisional countermeasures to help smooth out the problem-solving work. Naturally, provisional measures are only temporary even when specified by emergency response rules; they must be followed up with ongoing measures that deal with the source of the problem (Type B). A third approach (Type C) is quick investigation of the problem on the shop floor (a step beyond noting current conditions), followed by implementation of a solution.

The "Super Express" Strategy for Dealing with Rush Orders

Rush orders are one type of production problem to which this problem-solving system can be applied. They can have a wide impact on the production system for the following reasons:

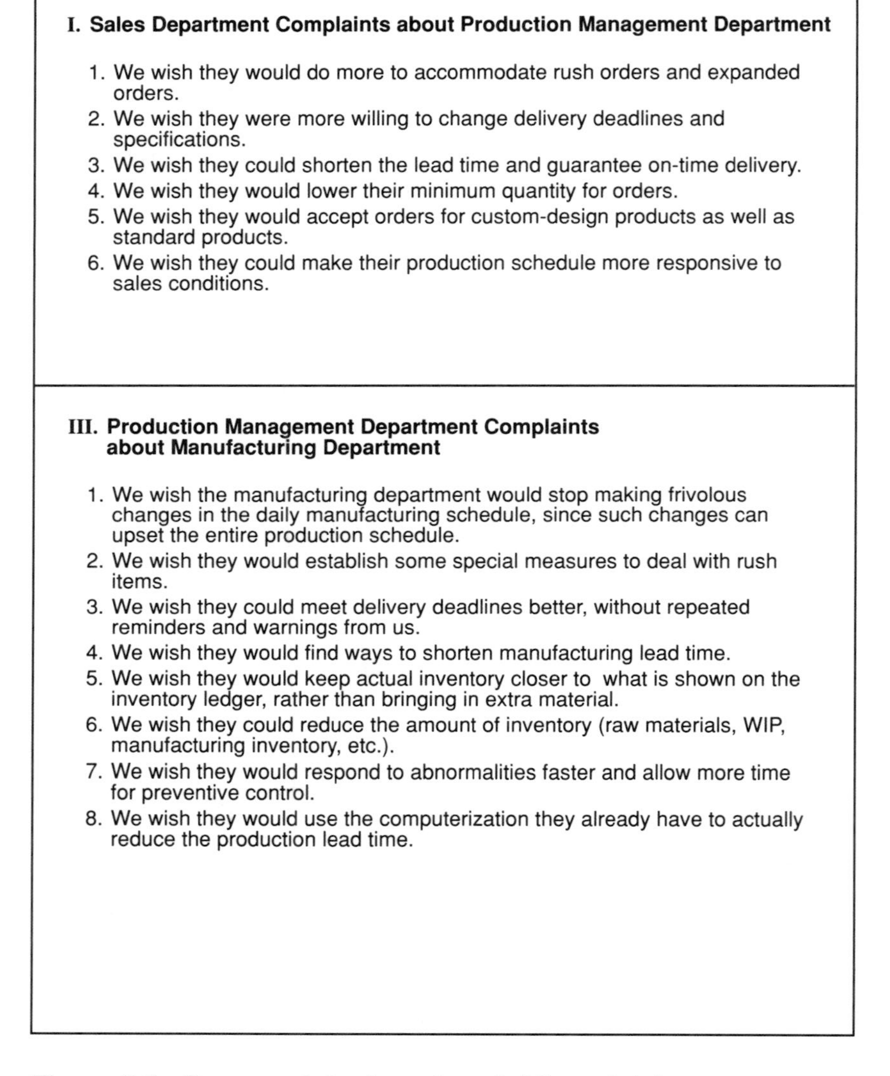

I. Sales Department Complaints about Production Management Department

1. We wish they would do more to accommodate rush orders and expanded orders.
2. We wish they were more willing to change delivery deadlines and specifications.
3. We wish they could shorten the lead time and guarantee on-time delivery.
4. We wish they would lower their minimum quantity for orders.
5. We wish they would accept orders for custom-design products as well as standard products.
6. We wish they could make their production schedule more responsive to sales conditions.

III. Production Management Department Complaints about Manufacturing Department

1. We wish the manufacturing department would stop making frivolous changes in the daily manufacturing schedule, since such changes can upset the entire production schedule.
2. We wish they would establish some special measures to deal with rush items.
3. We wish they could meet delivery deadlines better, without repeated reminders and warnings from us.
4. We wish they would find ways to shorten manufacturing lead time.
5. We wish they would keep actual inventory closer to what is shown on the inventory ledger, rather than bringing in extra material.
6. We wish they could reduce the amount of inventory (raw materials, WIP, manufacturing inventory, etc.).
7. We wish they would respond to abnormalities faster and allow more time for preventive control.
8. We wish they would use the computerization they already have to actually reduce the production lead time.

Figure 8-9. Common Interdepartmental Complaints

II. Production Management Department Complaints about Sales Department

1. We wish they would stop sending in rush orders and last-minute additions to previous orders.
2. We wish they would stop making changes in delivery deadlines and specifications.
3. We wish they would leave more leeway for delivery lead time when picking up orders.
4. We wish they would try to obtain more large orders.
5. We wish they would stop requesting custom-design products, so we can reduce variety.
6. We wish they would do more to avoid having to make changes in the production schedule.
7. We wish they would level out the influx of month-end orders.

IV. Manufacturing Department Complaints about Production Management Department

1. Frequent changes in the daily manufacturing schedule make it difficult to stick to the production schedule. The daily manufacturing schedule is separate from and not synchronized with the schedule of items purchased or supplied by outside manufacturers, causing frequent problems with delayed materials delivery and missing items — problems that lower our labor efficiency and equipment capacity utilization rates.
2. "Rush item" slips are sent to us so frequently that they cannot all be handled quickly.
3. We receive lots of warnings from the production management department, but no deadline extensions or cancellations.
4. The "paper lead time" is too long.
5. We cannot trust the inventory ledger. Some warehouse inventory exists only in the ledger and not in the warehouse, while some warehouse inventory is not registered in the ledger.
6. Instead of synchronizing the daily manufacturing schedule with the schedule of items purchased or supplied by outside manufacturers, they stick to the previous outside-item schedule, which means that some rush items come in slowly while some non-urgent items come in quickly.
7. We do not have a prediction-based management system (because we do not have early warning or quick-response systems), but we do give immediate feedback on problems. However, the production management shows no sign of making quick responses to our feedback.
8. We feel we can do a better, more accurate job by using our senses and intellect than be depending on a computer.

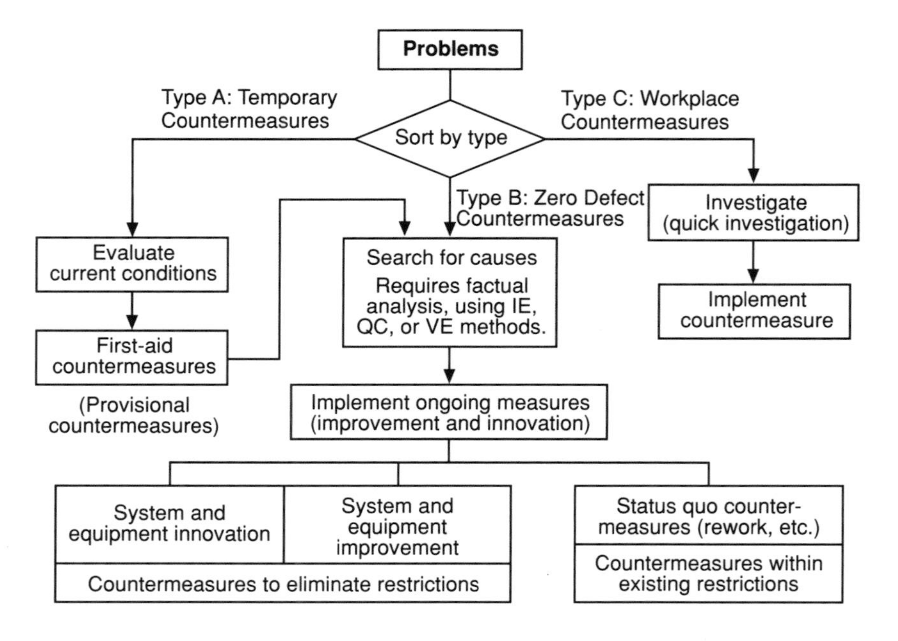

Figure 8-10. Different Countermeasures for Different Problems

- Rush orders can cause confusion at the work site because they interrupt production even during very busy periods.
- If information about rush orders is not communicated to everyone concerned, they can cause even greater confusion.
- They require managers to put in more time in following up on production progress reports (and sometimes require managers to make frequent progress reports to the customer).

Few manufacturing companies can afford to turn down rush orders, but most would prefer to avoid them. Here is how companies can systematize the way they handle rush orders.

Countermeasures for Type A Items (see Figure 8-10)

Study the conditions and implement a "first-aid" response.

1. Hold a meeting of everyone involved and plan out the daily schedule.
2. Make a special process chart (using enhancements such as color coding).
3. Use visual management methods.

Figure 8-11 shows an example of a rush order voucher. It has been given a "bullet train" motif, with a drawing and a catchy title to help take some of the tedium out of processing rush orders.

Because the rush order in this case is a prototype model, the voucher includes sections for quality evaluation marks and a time schedule. This slip is used as an item-list, work-instruction voucher that is sent along the production flow route with corresponding materials, as a sort of "kanban train" that stops at each station (process) where the materials are used. An enlarged version of this voucher is displayed as a visual chart so everyone can keep an eye on the progress of the rush order as it moves from process to process.

When we tried out this strategy at a factory, we found that people liked being able to check these "Super Express" kanban vouchers. We avoided the three problematic characteristics described above and instead established a smooth-flowing system for handling rush orders. The factory also received a letter of thanks from the rush order customer, and this letter was attached to the Super Express Strategy wall chart for everyone to read.

Countermeasures for Type B Items

After taking the "bullet train" to handle an emergency item, the line must have a plan for feeding the emergency item into the regular improvement schedule for ongoing management. We took the following measures:

1. We drew up a special notice to go with our monthly and daily production schedules.
2. We used color-coded paper to make the special notice easier to understand. Our special notice covered the following:
   - Items that exceed or might exceed quality standards
   - Rush orders or items noted as requiring high-priority control (including items not on today's schedule)
   - Prototypes, development items, test items, etc.
   - Equipment scheduled for repair work, etc.
3. We standardized the items in the special notice and composed a standard special notice form that can be output by the computer. Now this special notice can be used whenever the need arises, such as during production planning meetings.

Super Express Voucher

(Item list/work instruction voucher for rush item)

Special request: We need everyone's help to make the Super Express Strategy for this ____________________ a success!

Processes

Scheduled/actual times

Station (process)	Station A	Station B	—	—
Scheduled times	○○:○○	□□:□□	—	—
Actual times	:	:	—	—
Target/result	/	/	/	

Figure 8-11. Super Express Voucher for Rush Orders

4. The special notice may name extra preparatory activities, which should be treated as another process that needs to be controlled through standardization.
5. Items that require special attention are assigned to particular individuals who are responsible for thoroughly monitoring and managing those items.

These rules for preliminary investigation and organization of production management comprise an important ingredient in building a comprehensive management system.

The same systematic, rule-based approach should be taken when dealing with defective products or products that require rework. Naturally, it is better to focus on taking preventive action against defects instead of relying on post-facto measures. Any measures used for managing abnormalities are necessarily provisional measures.

KAIZEN FOR A SMOOTHER FLOW OF GOODS

This section describes improvement and standardization techniques for linking processes aimed at creating a smooth flow of goods among processes.

Managing the Information Links Between Processes

Production management systems need to meet various conditions imposed by the specific product and manufacturing situation; a system must be used in a way that is suitable to a particular factory. Figure 8-12 examines three types of production management systems based on factory conditions concerning the flow of goods. If you are creating or rebuilding a production management system, this figure will help determine which type of system your factory requires.

No matter which type of production management system you use, it undoubtedly needs some improvement. Use the levels shown in Figure 8-13 as a master plan for upgrading your system.

Improving Links Between Processes

Process flows vary a lot depending on the kinds of links that exist between them. To use another train analogy, we might note the differences between the

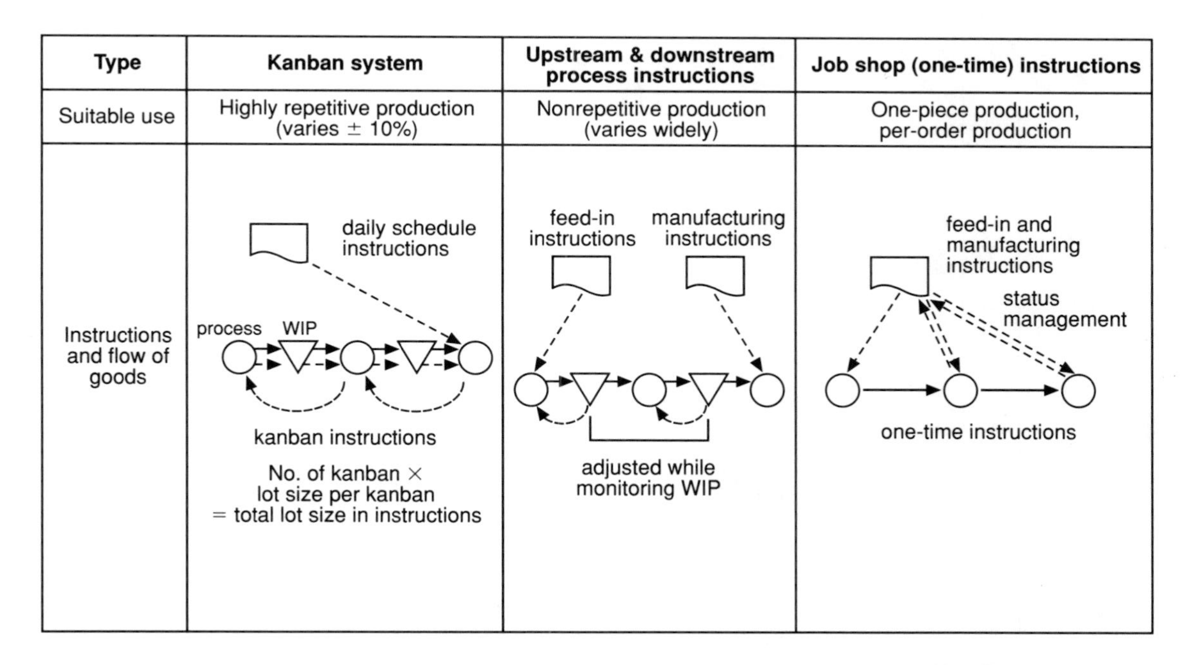

Type	Kanban system	Upstream & downstream process instructions	Job shop (one-time) instructions
Suitable use	Highly repetitive production (varies ± 10%)	Nonrepetitive production (varies widely)	One-piece production, per-order production
Instructions and flow of goods	daily schedule instructions process WIP kanban instructions No. of kanban × lot size per kanban = total lot size in instructions	feed-in instructions manufacturing instructions adjusted while monitoring WIP	feed-in and manufacturing instructions status management one-time instructions

Figure 8-12. Three Types of Production Management Systems Based on Flow Conditions

Element	LOW ←		Level	→	HIGH
1. Production system	Production per order; unregulated process flow	Use of GT approach for regulating process flow, workload dividend quantitatively	Systematic division of workload, leveled production	Small-lot production with frequent changeover	Automatic control, production with frequent changeover, FA/FMS
2. Layout	Machine-specific management	Combined flow, materials handling, rationalized linkages	JIT approach Level I	Level II	
3. Production management	Production based on monthly schedule; heavy dependence on worker skill and experience	Production based on weekly schedule, management by comparison of RT and ST	Production based on daily schedule, management by scientific ST method	Same as at left, but with hourly control based on short cycle times	Second-based control using automated control systems
4. Equipment management	PM practiced by maintenance department	Introduction of companywide PM (TPM)	Establishment of preventive maintenance	Establishment of improvement maintenance	Implementation of maintenance prevention
5. Quality control	Quality review once per month	TQC activities, small-group activities	Mistake-proofing activities, quality control at the source	Building in quality via high equipment quality	Overall production linear flow rate of 98% or higher*
6. Equipment automation	One operator, one unit	Multiunit and multiprocess handling	Unassisted operation within normal operating hours	Automatic operations during operating hours and break periods	Computer-controlled operation
7. Administrative system	Manual batch processing	Partial mechanization of manual work	Automation of usual tasks	Total online system	Coupling of automated equipment and clerical processing equipment
8. Work-in-process management	Left to worksite managers to decide	Standards for storage sites and amounts stored	Countermeasures to reveal latent WIP-related problems	Devices to enable level production with minimal WIP	FA/FMS includes management of WIP and parts warehouse

*Linear flow means production moves in a smooth, trouble-free flow, without additional process loops such as rework.

Figure 8-13. Five Improvement Levels for Main Elements of Production Management

travel time of express and local trains: You can get from Station A to Station B in 25 minutes on an express train, whereas it takes 45 minutes to cover the same route on an ordinary train. The time difference results from the ordinary train going more slowly and stopping at every station between Stations A and B.

In factories, the little stops that lengthen the travel time of goods between processes are called *storage points*. One lot of workpieces sits at Process A until the entire lot is finished, then it moves to the next process. The longer the time required to work on each workpiece at each process, the longer the lot of workpieces must await conveyance to the next process.

Figure 8-14 shows the time amounts taken by various steps in a foundry's production, up to the packaging process. These pie charts clearly depict the large amount of storage time. Based on this analysis, the foundry managers took a cold, hard look at the time spent doing everything other than net production time, since only net production time adds value to products. They began planning measures to drastically reduce storage time, manual handling, and conveyance. Their measures contained three key points:

1. They raised the transport readiness index for storage sites. (The transport readiness index indicates the ease or difficulty involved in conveying and handling items; it considers such factors as the types of objects to move and the way they have been placed.)

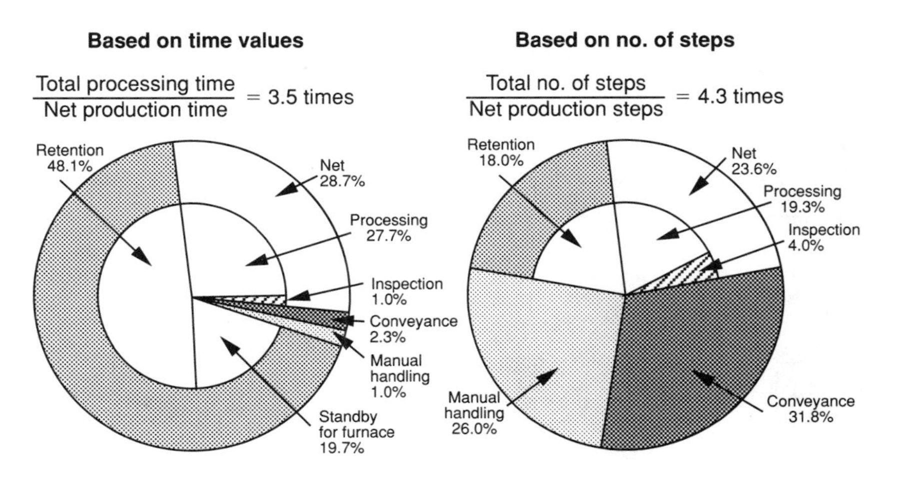

Figure 8-14. The Significance of Retention Time (Foundry Operations Process Analysis Example)

2. They improved the links between processes, as shown in Figure 8-15.
3. They streamlined the standard times for each process (by making time-saving improvements) and thereby reduced processing times.

To make effective improvements in the process flow, you need analytical methods such as these, as well as systematic measures. The reason being that the changes you make must produce a layout that works immediately and that is worth the money invested in the improvement.

How Changeover Creates Work-in-Process

I would like to return briefly to the subject of changeover, this time as it relates to production management and specifically to the issues of economic lot sizes and work-in-process.

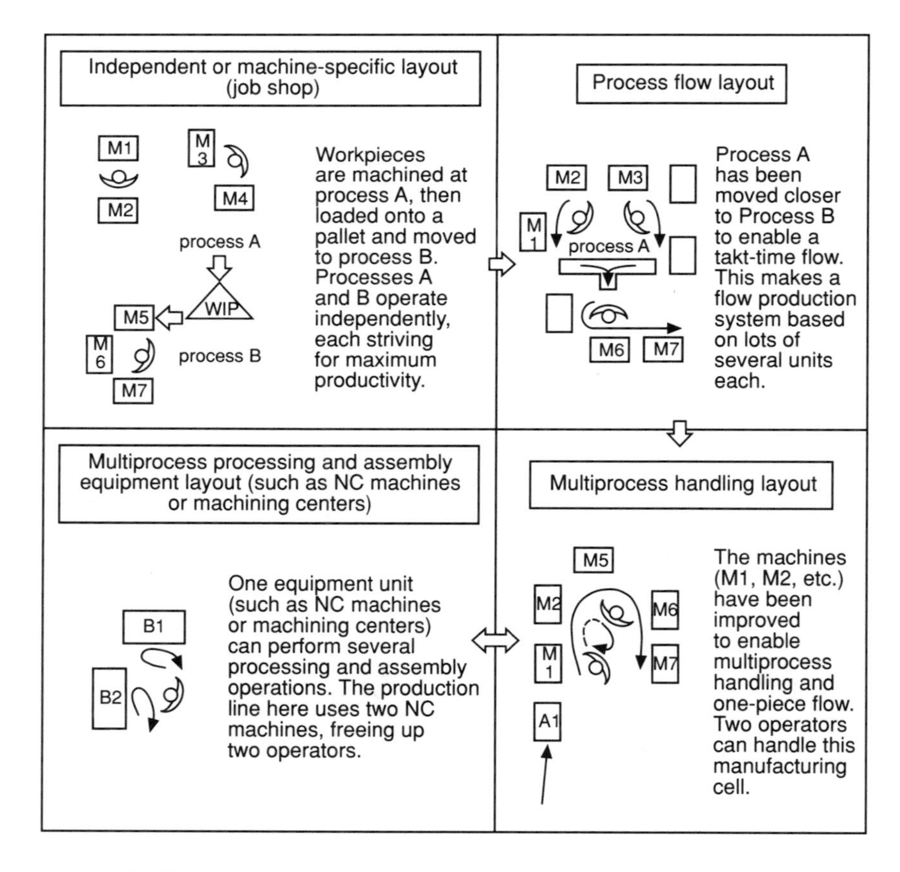

Figure 8-15. Process Link Improvements

The diagram in Figure 8-16 shows how a key developer of the economic lot theory, Dr. Shigeo Shingo, defines economic lot size. (He was introduced earlier as the inventor of quick changeover approach.) This economic lot theory, based on the relation between cost and lot size, shows that larger lots can be made less expensively, but inventory costs eventually exceed production costs. The economic lot size is the balance point between production costs and inventory costs.

Now look at Figure 8-17. Given a changeover time of eight hours and a lot size ranging from 100 units to 10,000 units, changeover has a much smaller impact on operation time when the lots are large. When the operation takes one minute per piece and changeover takes eight hours, 480 units of work-in-process stack up during a changeover.

What if you could reduce the changeover time from eight hours to just five minutes? If you process 100 units per hour, the operation time per hour would be 1 minute + (5 minutes ÷ 100) = 1.05 minutes per unit (it would be for 1.005 minutes per unit at 1,000 units per hour and 1.0005 minutes per unit at 10,000

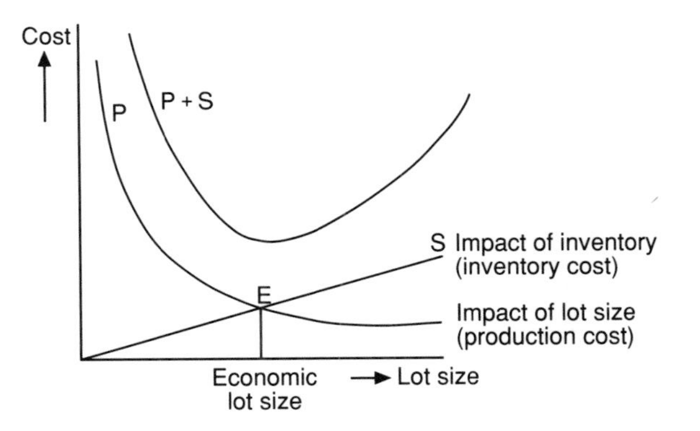

Figure 8-16. Original Theory of Economic Lot Size

Changeover time (hours)	Lot size (units)	Net operation time/unit	Total operation time/unit including changeover	Changeover time ratio*
8	100	1 min.	1 min. + $\frac{8\times60}{100}$ = 5.8 min.	82.8%
8	1,000	1 min.	1 min. + $\frac{8\times60}{1000}$ = 1.48 min.	32.4%
8	10,000	1 min.	1 min. + $\frac{8\times60}{10,000}$ = 1.048 min.	4.5%

*Changeover time per unit divided by total operation time per unit.

Figure 8-17. Relationship between Changeover Time and Lot Size

units per hour). This means the amount of work-in-process created would be reduced from 480 units to approximately 5 units, no matter the size of the lot.

Once you reached this point, you could reduce the lot size to 5 or 10 units and have very little work-in-process and no waiting time for changeover. This is the objective of quick changeover. When you achieve it, you clear the way for wide-variety, small-lot, mixed-flow production. Changeover improvement can indeed have a large effect.

Toward Overall Flow Improvement

By adding quality-oriented countermeasures to the types of countermeasures just described, you can implement a comprehensive approach to flow improvement that overlooks nothing. Such comprehensive flow improvement helps reduce the amount of work-in-process and makes for a smoother flow of goods. This kind of approach is best pursued step by step.

Sometimes, however, such improvements must begin in situations where the current layout is bad or where other basic problems must first be corrected by process improvements. In such situations, the first step must be to make a detailed analysis of the current conditions, following a step-by-step plan, such as that shown in Figure 8-18. This SN type of plan includes detailed steps and can be very effective.

The basic approach behind this plan includes the following points:

1. A clear, clean flow makes management simpler and eliminates the need for complex computer processing. Factories should try to develop this kind of flow.
2. To make processes more efficient, you need to establish a cycle time, improve links between processes, and develop better techniques, equipment, and work sequences that build quality into products at each process.
3. Various production management methods control the what, where, when, and how variables of production.

Figure 8-19 outlines a production management improvement plan that is a further development of the improvement levels shown in Figure 8-13. This plan points to the need for comprehensive improvement of the production management system.

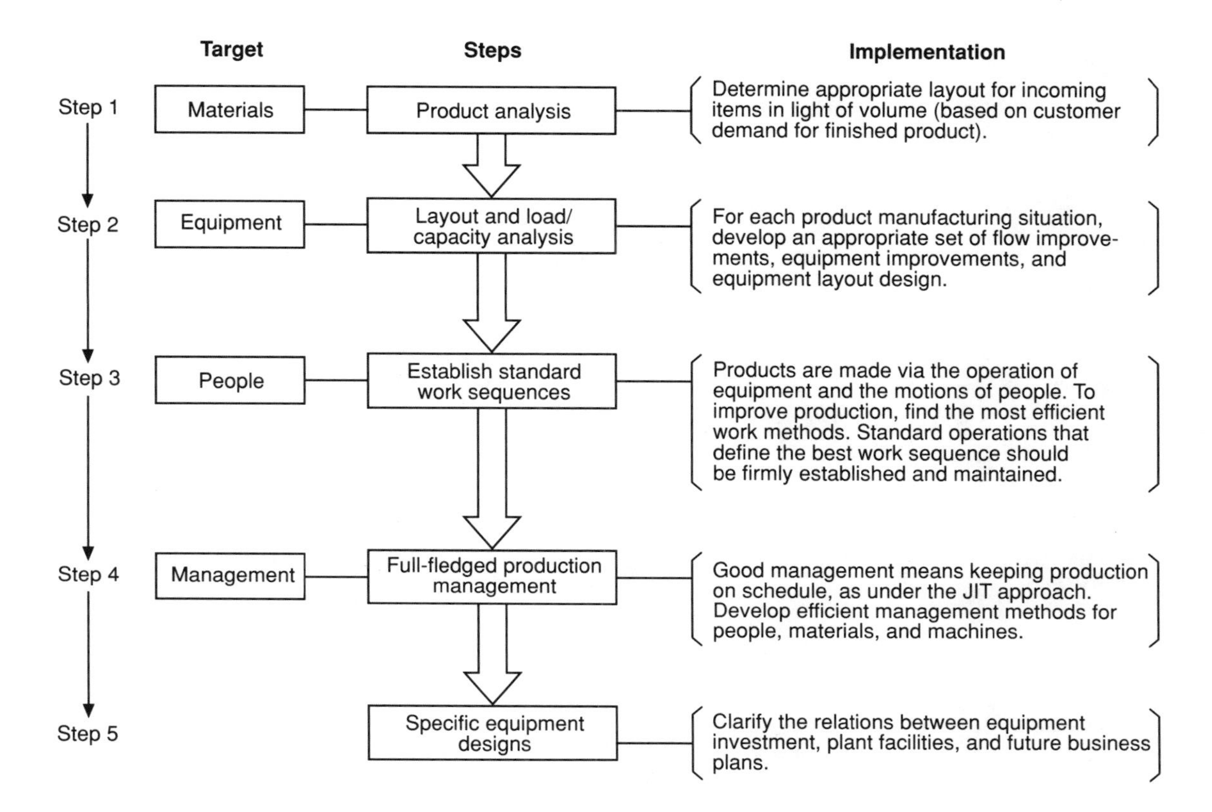

Figure 8-18. Steps in Layout Design (SN Approach)

In reality, FA (factory automation) and FMS (flexible manufacturing system) are not types of production management systems but rather result from a standardization approach that has upgraded the various production management system elements listed in Figure 8-13 to a very high level. The goal is not to achieve FA or FMS as an end in itself, but to implement these kinds of improvements and standardization.

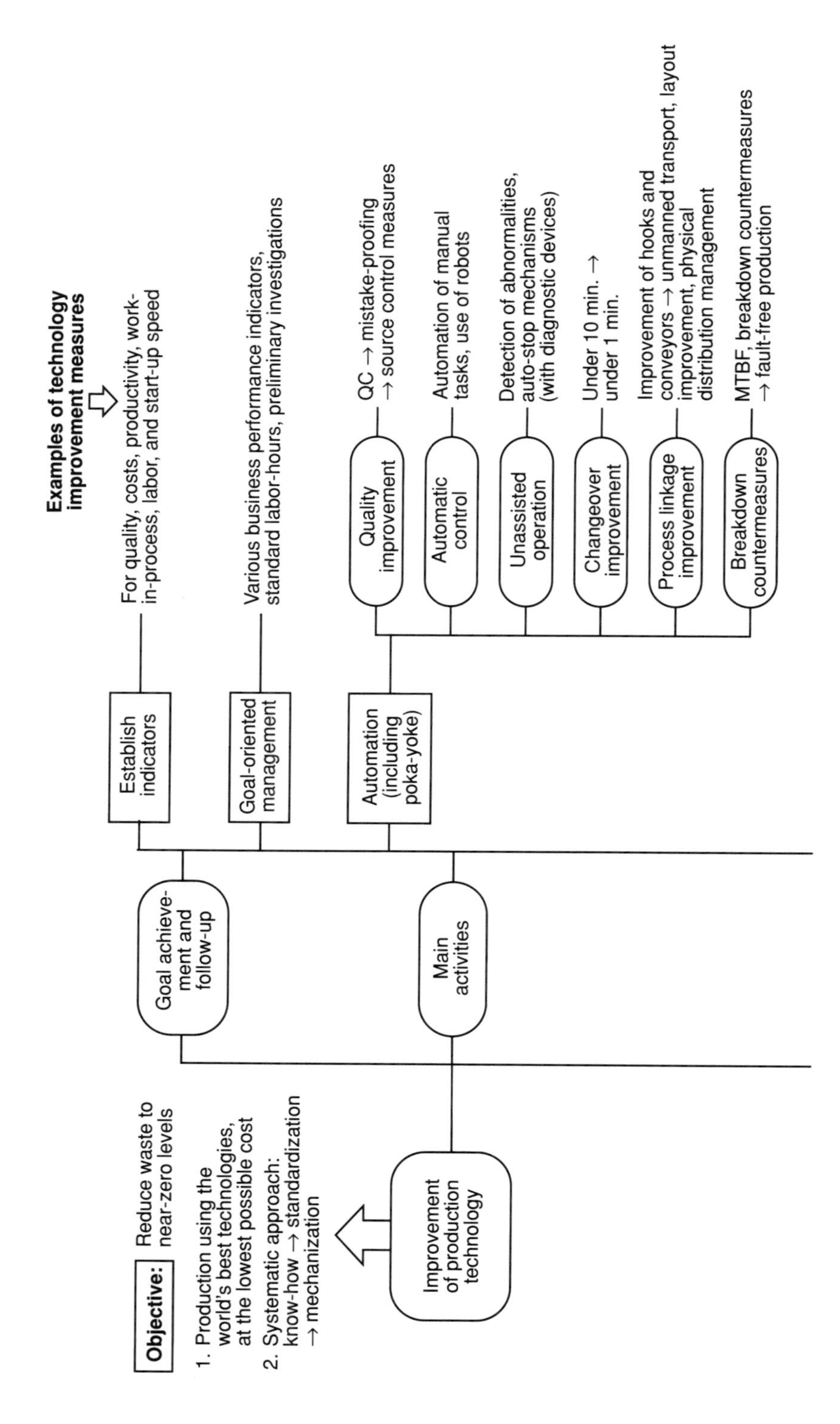
Objective:
Reduce waste to near-zero levels
1. Production using the world's best technologies, at the lowest possible cost
2. Systematic approach: know-how → standardization → mechanization
Improvement of production technology
Goal achieve-ment and follow-up
Main activities
Establish indicators
Goal-oriented management
Automation (including poka-yoke)
Examples of technology improvement measures
For quality, costs, productivity, work-in-process, labor, and start-up speed
Various business performance indicators, standard labor-hours, preliminary investigations
Quality improvement
QC → mistake-proofing → source control measures
Automatic control
Automation of manual tasks, use of robots
Unassisted operation
Detection of abnormalities, auto-stop mechanisms (with diagnostic devices)
Changeover improvement
Under 10 min. → under 1 min.
Process linkage improvement
Improvement of hooks and conveyors → unmanned transport, layout improvement, physical distribution management
Breakdown countermeasures
MTBF, breakdown countermeasures → fault-free production

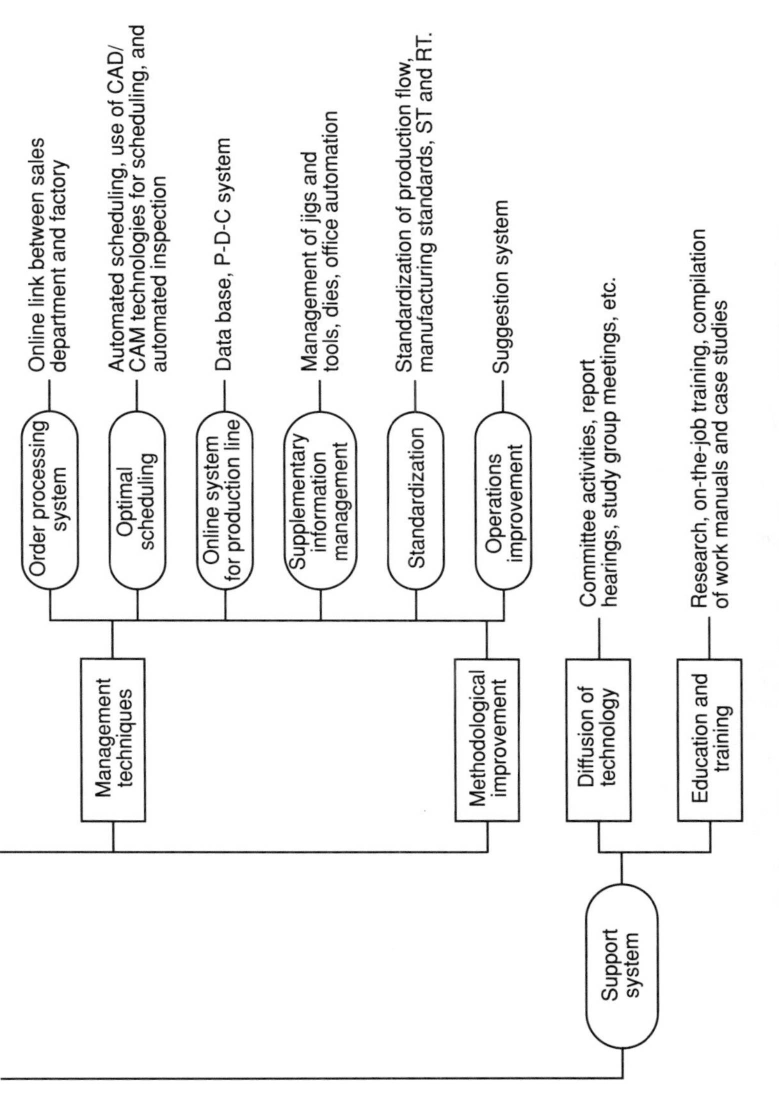

Figure 8-19. Plan for Upgrading the Production Management System to Become an FA/CIM System

9
Standardization for Decision Making

There are three types of problems that involve decision making:

1. Problems that require breakthrough innovations.
2. Problems that apply the accumulated skills and experience of the employees.
3. Problems that can be resolved through standard procedures, which are often automated.

The highest-level problems belong to the first category; these are often addressed with techniques for generating creative ideas. The third category includes problem-solving routines that can be automated, some routines being more difficult to automate than others. With the ongoing progress in artificial intelligence, more of these routines will be automated in the future. Although these subjects are interesting, they are beyond the scope of this discussion.

This chapter discusses the second kind of problem solving, where people contribute directly based on their own experiences. We will examine three examples in which standardization has been used for this kind of decision making.

STANDARDIZATION OF MEETING PREPARATION

The IE, QC, and VE techniques described earlier can be used effectively in project-team efforts to solve problems. Consider how these techniques can apply to problem solving in the context of the kinds of meetings typically held at factories.

Figure 9-1 outlines the general functions of such meetings and the kinds of standards that help ensure success. If you omit any of the three items listed under the "input" section of the figure, your meetings will take longer and have less chance of reaching successful conclusions. Opening a meeting that does not benefit from these kinds of "input" preparations is like opening a factory that has substandard parts and materials, unskilled workers, and constantly malfunctioning equipment.

As for the "action" section, clearly defined objectives, scheduling, and visual aids (such as drawing on the board, or bringing actual parts to discuss) keep the meeting on track. Figure 9-2 describes more specifically the actions recommended for approaches (a) to (e) under the "action" section.

As described under "output," the conclusions of such meetings should be thoroughly reviewed to make sure nothing was overlooked and to revisit the key

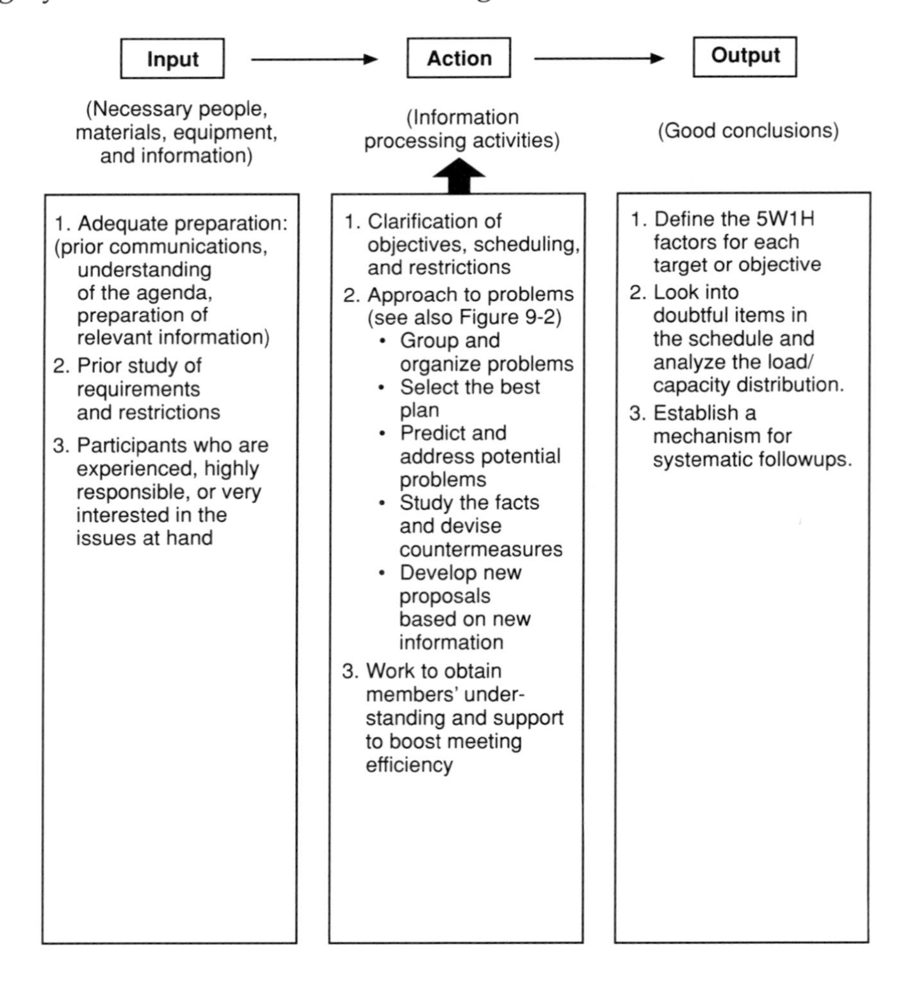

Figure 9-1. How to Make Meetings Function More Smoothly

Approach	Elements of countermeasure
(a) Organize and prioritize the problems to be addressed	1. Write problem themes, such as "What position should we take in dealing with XX?" 2. List all items of concern for each theme. 3. Rank themes according to priority, from "very important" to "not pertinent." 4. List examples, wishes, and requirements concerning the more important themes. 5. Evaluate the overall importance of items listed in (4). 6. Apply the 5W1H approach in dealing with the most important and most urgent themes.
(b) Select the best plan among all those proposed	1. Write up the objectives targets, and conditions as themes. 2. List the evaluation criteria. Sort items such as specifications, advantages, disadvantages, and desired items into"musts" and "preferences." (See Chapter 2). 3. Make up a matrix chart listing the various ideas across the top and the evaluation criteria, in priority order, along the sides ("musts" first). 4. Use a scoring system or symbols for the evaluation, and attach notes explaining reasons. 5. Select a proposal. 6. List the selected proposal's drawbacks and draft a counterproposal. Note: It may be helpful to evalute several competing proposals, such as layout proposals for linear, L-shaped, or U-shaped cells.
(c) Look into issues that may become problems in the future.	1. Use the 5W1H format in describing each theme's objectives. 2. Draft a daily schedule. 3. List doubtful items and explain why they are doubtful. 4. Draft solution proposals. Focus on important problems and determine by whom, by when, and how each problem is to be solved. 5. Report on the solution and conduct a followup (such as for delivery management or relocation of equipment).
(d) Investigate the facts related to each problem and devise counter-measures.	1. Identify the themes and study their scope. 2. Go to the worksite and take a hands-on approach, using the 5W1H method for probing the facts. 3. Draft countermeasure proposals at the worksite, then discuss them at the meeting and apply the 5W1H method to organize their contents.
(e) Develop new proposals based on new information:	1. Rewrite the project themes. Hold an initial brainstorming session. Use the methods described in (d) to obtain any other required information. 2. Conduct any necessary external information research (e.g., market, technical, patent) and organize results by function. Study other lines in the same factory. 3. Use the methods described in (b) to select a proposal for implementation.

Figure 9-2. SN Approach to Problem Solving

points discussed. If possible, the decisions and other conclusions reached at the meeting should be typed and distributed to the meeting participants they depart.

Good conclusions include steps for following up on the effects of the implemented activities. These are the D and C elements of the P-D-C cycle. Figure 9-3 shows how to standardize preliminary steps that help ensure a good conclusion.

STANDARDIZED PROCEDURES FOR RESPONDING TO COMPLAINTS

In one respect, products using technologies that are not yet fully developed cause customer complaints; in another respect, however, these products provide

people to be contacted: meeting title:
Facilitator:
Notice of Meeting
Date:
Meeting topic: | Site: | Time: ___ to ___
Attendees:
1. Objective
2. Theme (and relation to past themes, if any)
3. Items and presenters
status of preparations
Theme

To: ___ **Minutes** Next meeting date:
Meeting title: | Site: | Time: ___ to ___
Attendees:
Absent:
1. Objective:
2. Matters discussed

No.	Theme	Matters discussed and conclusions	To be implemented by

3. Conclusions
4. Themes and issues for next meeting
5. Additional comments

Keys to Effective Meetings

Preparation: Define and think through the intended result and prepare thoroughly.

Attendance: Select only those participants needed to achieve the purpose at hand. In principle, only one person from each section, the optimum for an action-oriented meeting is six to nine people.

Communication: Make sure all the necessary people can attend and maintain close communications. Schedule the meeting after deciding who should come.

Documentation: Meeting notices should make clear the objective, provide a detailed agenda, and indicate the status of preparations. Meeting minutes should succinctly document the discussion and indicate when action items are to be carried out and by whom, they should be accompanied by a notice for the next meeting.

Timing: Emphasize preparation and begin and end on time. One-hour meetings are best, two-hour meetings are good, three-hour meetings should be the maximum allowed. Keep Wednesday a no-meeting day.

Facilitation: Leaders should agree on the agenda and other procedural matters in advance. Do not stray from the objective; keep discussion on track, but discuss matters thoroughly. Avoid win-lose situations and discourage opinionated or self-serving arguments. Work for consensus, not just majority rule. If a meeting lacks required information or a key person, adjourn and reschedule.

Figure 9-3. Standardization of Preliminary Steps for Meetings

opportunities to discover key points for technology improvements. Complaints indicate quality differences among competitors; sometimes a customer includes complaints as elements of a generally constructive proposal to the manufacturer. In many cases, these complaints have planted the seed for the development of new products and technologies.

Even when manufacturers are certain they have tested their products under every condition imaginable, their clients manage to use them in unexpected ways and run into problems. Nevertheless, manufacturers should welcome complaints from customers and explicitly thank the customers for their cooperation.

Complaints should be met with more than empty talk at management meetings. They demand a practical factory-based response. In standardizing complaint-handling procedures, the objective is the same as in other types of standardization: to make the procedures easier, quicker, and more reliable. Consider the following example of one way to handle customer complaints. One day, a service clerk received a call from an angry customer: "That product you sent me is a mess! I'm not paying for it. Come take it back!" The customer went on to explain that the product had arrived in a broken box, which meant that it could not be shipped on to the end user by the promised date. This angered the customer because he was afraid of losing the end user's confidence.

Clearly, this situation called for immediate "first-aid" measures like the ones outlined in Figure 8-12. The clerk responded: "Listen, we'll send someone right over to repackage the product, and meanwhile we'll find out if we've got an identical product in stock. If so, we'll send it right away." The clerk promised to call back within three minutes to confirm that someone was coming to repackage the product, then hung up. Three minutes later, he called the customer to say that their engineer was on the way would arrive at about X o'clock.

While on the phone this time, the clerk picked up a customer complaint form (see Figure 9-4) and obtained the basic facts. He took the completed form and used it as an urgent action request, sending copies to several relevant people.

Hands-on implementation is essential. In this case, dwelling on opinions, assumptions, theories, or ideas for countermeasures would have been useless. The engineer who received the urgent action request from the clerk went to the factory to determine the cause of the faulty packaging. This was important not only to find the cause but also to explain it to the customer and to take countermeasures to prevent the problem from recurring. Without these steps, the manufacturer would find it more difficult to obtain new orders from the customer and could lose the customer to a competitor.

Complaint Report Form

To:________________ Date:________

Manufacturer:________________ Countermeasure required: YES NO

Route to:	Client code:____ Form No.:______	Issuer	
Part name/model			
Customer and delivery destination		Type of complaint	
Volume and value		Date of complaint	

1. Description of problem (use 5W1H method)
2. Customer's and sales agent's comments
3. Confidence damage appraisal

Complaint Countermeasure Form

Date issued:________

	Form No.____	Office code:______
Customer and delivery destination		
Item		Branch code:______
Specifications		

Amount delivered		Manufac-ture date:		Date of complaint:	
Damage reported by customer				Severity rating:	

1. Description of complaint and cost of damage
2. Cause of complaint
3. Investigation of cause (5 whys)
4. Countermeasure and communication with relevant parties
5. Countermeasure for delivered items
6. Customer's comments
7. Confidence damage appraisal
8. Appended documents

Figure 9-4. Complaint Report Form and Complaint Countermeasure Form

Next, the manufacturer obtained the customer's cooperation in reproducing the cause, which in this case was traced to the customer's receiving dock operations. Through investigation and testing, they found that the damage was caused at the customer's dock by new forklift prongs that have a different shape from the old prongs. The new prongs tended to catch inside the metal bands on the product boxes. The manufacturer's engineer checked all of the customer's forklifts and how they carried the boxes. At the same time, he wrote a report on the problem and worked out some countermeasures with the customer. He left the customer with a promise to send a copy of the finished countermeasure plan by a certain date.

Back at the manufacturing company, the people involved assembled the required materials and held a meeting. They decided on a countermeasure: changing the pallets and metal bands to prevent the forklift prongs from causing damage. After doing a thorough study of the relevant conditions at the receiving dock, they standardized their improvements. They sent a copy of the countermeasure plan (see form in Figure 9-4) to the customer and obtained the customer's understanding regarding their improvements. These same basic steps can be used when processing almost any kind of complaint.

Sometimes, this factory-based approach does not work well enough. In such cases, the manufacturer may have to write a new manual, conduct new training courses for its repair staff (and sometimes for the customer), or even establish a specialized repair service. In these cases, standardization should also be implemented at the highest possible level. Here we should refer back to the SN evaluation chart (shown in Figure 3-4), which describes the improvement route:

Men/Women (work force) → Methods → Jigs, tools, measuring devices → Mechanization → Materials

People's TV sets, washing machines, and other appliances used to malfunction frequently, keeping neighborhood repair shops rather busy. These days, such products rarely break down. In terms of standardization levels, this better situation exemplifies methodological improvements.

THOROUGH REVIEW OF PLANT INVESTMENT PLANS

Let us now consider the objective of reliable, easy, and quick planning for major plant investments, looking at how to apply standardization to this important function. One approach establishes standard check items to prevent omissions and oversights. These items are written up as a checklist for evaluating key points at each stage. Such a checklist can be very effective.

Figure 9-5 shows a three-stage organization of plant investment plans. By applying the SN approaches shown in Figure 9-2 to these four stages, we can do a thorough check of the plans. It would also be useful to apply the checkpoints shown in Figure 9-6. These checkpoints are just the basic elements; you will need to adapt them to fit your company's conditions.

Next, you will compile the results of your planning investigations and enter them into the investment plan file using forms such as those shown in Figures 9-7 and 9-8. Figure 9-7 provides a comprehensive outline used by administrative workers, while Figure 9-8 includes standardized overall evaluation points for managers.

After completing these steps and finalizing the plan, you still need to standardize the handling of revisions that sometimes must be made when conditions change during implementation. Figure 9-9 provides a form for listing predicted variables and describing the countermeasures planned for each of them. This type of form can be useful when preparing for such contingencies.

The standardized measures described above prepare you for the worst, since the best takes care of itself. As such, these measures help make your planning work easier, faster, and more reliable.

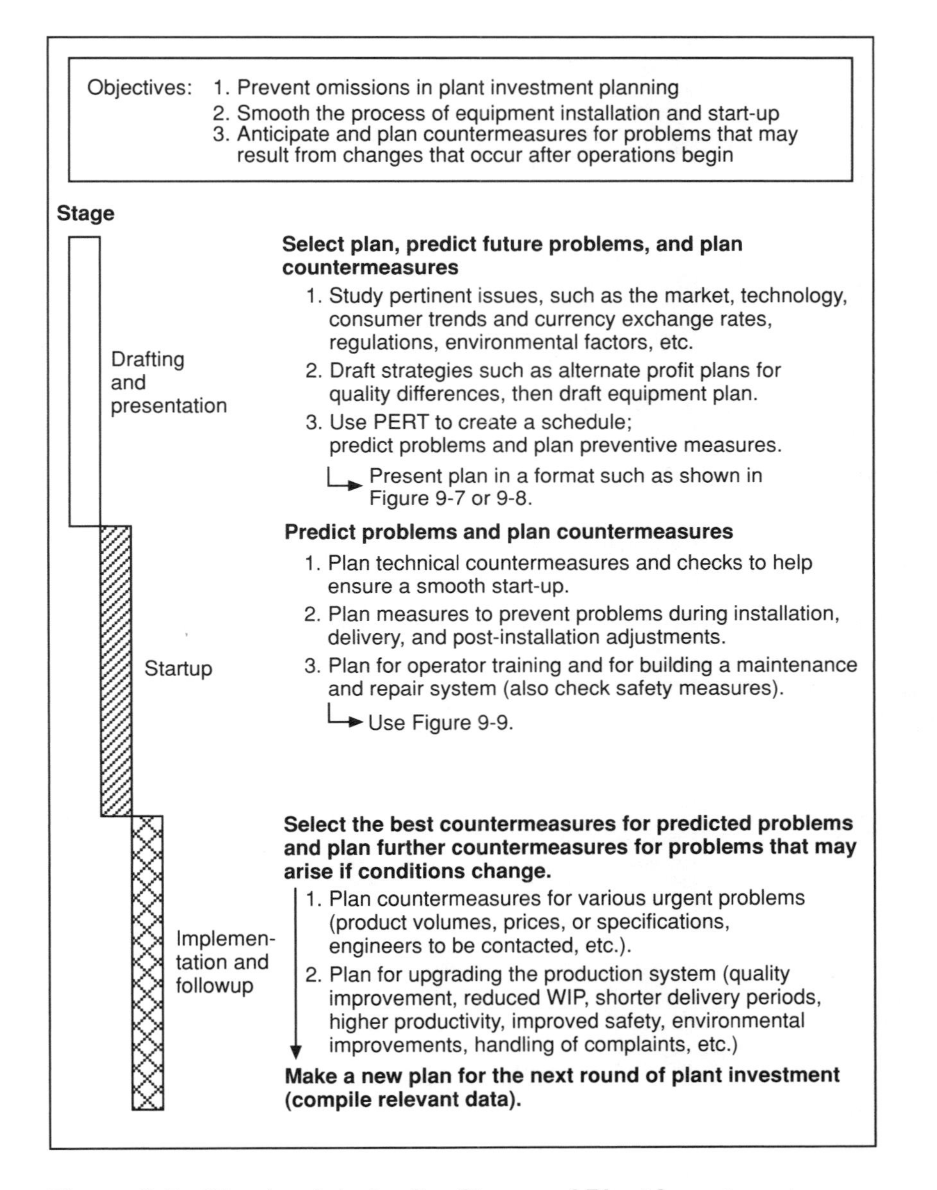

Figure 9-5. Checkpoints for the Stages of Plant Investment Planning

Category	Item	Check	Counter-measure
Market Factors	1. Market size (current and predicted) 2. Market share, trends 3. Products with competing materials/functions 4. Sales strength (customer share, future potential) 5. Price trends (market prices) 6. Quality competitiveness 7. Price competitiveness 8. Product life 9. Other (specify:______________)		
Planning	1. Clarification and reliability confirmation of sales period, sales volume, and sales prices 2. Clarification and reliability confirmation of rationalization objectives (staffing reduction, flow of goods, reduced WIP, quality improvement, absorption of subcontracted work, energy conservation, etc.). 3. Technical achievement levels (process reduction, robotization, computerization, changes in production methods, policies, or materials used, etc.) 4. Other (specify:______________)		
Plant and Equipment	1. Equipment comparisons (with existing equipment, with similar equipment, with competitors' equipment, etc.) 2. Use factors (ease of use, design customization possibilities, load vs. capacity, maintainability, required tools, wear and tear, breakdown rate, parts compatibility, etc.) 3. General-purpose or specialized (contingencies concerning design changes, policy changes, etc.) 4. Environmental measures, safety measures 5. Changeover time 6. Suitability of delivery period, activities after delivery (installation, test operation, etc.), after-sales service 7. Other (specify:______________)		
Profit Planning	1. Countermeasures to safeguard profit plan (including rationalization measures if output is being reduced) 2. Profitability impact of shorter production and delivery lead times and reduced in-process inventory 3. Accuracy and reliability of profit plan 4. Need for additional investment to cover higher running costs following new equipment installation 5. Connection to future cost-reduction measures 6. Other (specify:______________)		
Other	1. Set scope of current-condition studies (product analysis, production flow analysis, load estimation, assignments and job descriptions) 2. Study further simplification and automation of production management 3. Study relevant patents, regulations, etc. 4. Review studies into additional facilities and equipment. Check site conditions. 5. Other (specify:______________)		

Note: Use "Check" column to enter priority rank numbers.

Figure 9-6. Checklist for Plant Investment Planning

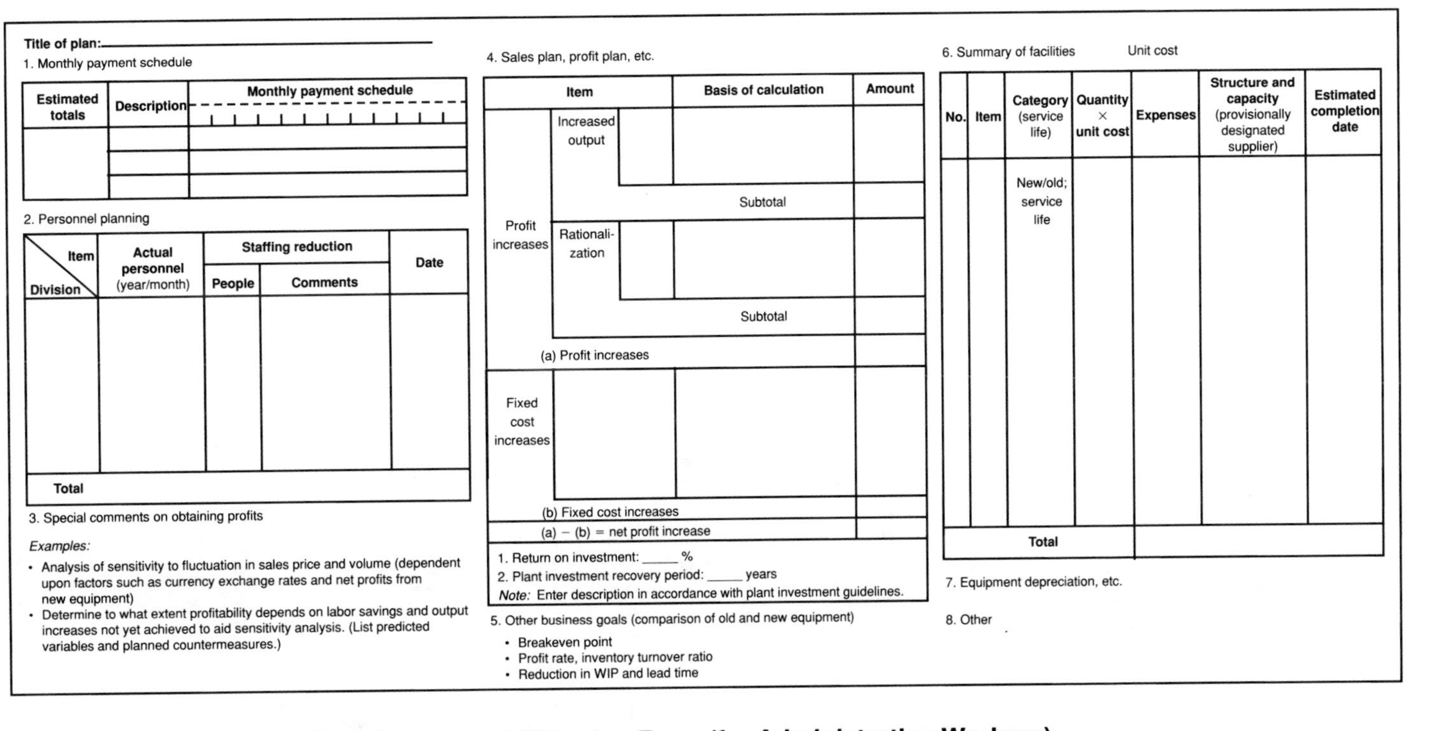

Title of plan: ____________

1. Monthly payment schedule

Estimated totals	Description	Monthly payment schedule

2. Personnel planning

Item / Division	Actual personnel (year/month)	Staffing reduction		Date
		People	Comments	
Total				

3. Special comments on obtaining profits

Examples:

- Analysis of sensitivity to fluctuation in sales price and volume (dependent upon factors such as currency exchange rates and net profits from new equipment)
- Determine to what extent profitability depends on labor savings and output increases not yet achieved to aid sensitivity analysis. (List predicted variables and planned countermeasures.)

4. Sales plan, profit plan, etc.

Item			Basis of calculation	Amount
Profit increases	Increased output			
			Subtotal	
	Rationali-zation			
			Subtotal	
(a) Profit increases				
Fixed cost increases				
(b) Fixed cost increases				
(a) − (b) = net profit increase				

1. Return on investment: _____ %
2. Plant investment recovery period: _____ years

Note: Enter description in accordance with plant investment guidelines.

5. Other business goals (comparison of old and new equipment)

- Breakeven point
- Profit rate, inventory turnover ratio
- Reduction in WIP and lead time

6. Summary of facilities Unit cost

No.	Item	Category (service life)	Quantity × unit cost	Expenses	Structure and capacity (provisionally designated supplier)	Estimated completion date
		New/old; service life				
		Total				

7. Equipment depreciation, etc.

8. Other

Figure 9-7. Example of Plant Investment Planning Form (for Administrative Workers)

Title of plan: ______________________

Sponsor:	Division:	Sales planning mgr.:		Division mgr.	Plant mgr.
		Production planning mgr.:			

Objectives

1. Background (what the plan is intended for)
2. Objectives (what? how?)
3. Plan summary (specify methods to meet objective)

Goals

1. Plan summary

Total investment cost

Measures for sales, labor efficiency, and rationalization

Item	Before (year/month)	After (year/month)	Comments

2. Profit plan (attach notes showing formulas for sales increase and cost reduction
   - Return on investment: _____ %
   - Plant investment recovery period: _____ years
   - Break-even point, etc.
3. Other (reduction of WIP, etc.)

Summary of Investment Plan

Summary example

1. Description of current conditions and items targeted by plant investment plan (such as sales figures, market share figures, sales plans)
2. Summary of conditions before and after improvement and key improvement points
3. Charts and diagrams to describe the above

   List charts and diagrams for items not shown here; attach copies.

Examples: Flow diagrams, equipment sites, layout diagrams, environmental improvement diagrams, product manufacturing process outlines, photographs, CPU system outline.

Figure 9-8. Example of Plant Investment Planning Form (for Managers)

Outline of Facilities

1. Procured equipment, etc. Total cost: _______ (Supplemental cost: _______)

No.	Name	Units	Cost	Comments (specifications, capacity, etc.)
	Enter only main items; enter other items on subsequent pages			
	Plant investment cost			Capital procurement method
	Supplemental cost			
	Total plant investment cost			

2. Equipment depreciation (main items only)

Planning and Construction Schedule

Priority ranking: __________ Construction completion date: __________

Operation start date: __________

Item	Schedule (year/month)

Countermeasures for Predicted Problems

	Risk items	Who	Period	Countermeasure
	Examples 1. Sales volume/prices trends 2. Response to competing technologies 3. Trends among competitors 4. Behind-schedule situations 5. In-house technology level 6. Profit planning measures			

Special comments, appended documents, etc.

Project team name: ______ Investigation date: ______ Implementation start date: ______

1. Objective: Project team ____ will plan countermeasures to help Plant Investment Plan ____ meet its goal of ____ (enter target) by ____ (date).

Problems and conditions (describe using 5W1H)	Investigation of causes (using "five whys" approach)	Eval.	Planned countermeasures (including preventive measures)	Eval.	Implementation schedule			
		Contribution to goal (%)		Effect	Who	Items	Schedule	
							Period	Check
1. Describe problem points and examples 2. Rank importance from 1 to 5. If problem has a high probability of occurring, also enter its risk level (1 to 5).	1. List with reference to checklist 2. Write neatly and without omissions. 3. Ask "why?" repeatedly to get to the root of the matter. *Example:* A competing product arrived unexpectedly. ↓ Why? Product was weak. ↓ Why? It did not fully meet customer needs.	A B	1. Develop several alternative plans based on Theme A. Evaluate each plan. • Use same methods as Company M. • Plan to introduce new technology. • Introduce and apply XXX technology.	A B (A)	— — MK	• Test new product XXX. • Obtain approval from client.	M/D	✓

Risk level

1 Not significant with very low risk
2 Significant, but low risk
3 Significant, with 50/50 risk
4 Significant, with high risk
5 Significant and certain to happen

Further investigation (stamp here)

(A) Absolutely required
A Required
B Recommended
C Not required for now, or handled by another improvement plan

Contribution level (Circle indicates countermeasure selected)

A Is very easy to do and will be very effective.
B Requires a little effort and will be very effective.
C Requires some effort, but will be effective.
D Requires much effort, but is feasible. May not be very effective but is still desirable.
E Requires much effort, not very feasible, not likely to be effective

Figure 9-9. Plant Investment Plan Potential Problem Form

10
How Standardization Can Help U.S. Operations

From 1988 to 1990, I lived and worked in the United States as a production work-site manager, helping to build and start up a Japanese company's U.S. production facilities. These facilities included casting, processing, painting, and inspection processes and employed a work force of about 200.

This project, which brought together many new employees, presented me with many important process and equipment management responsibilities with regard to developing production management and process management systems. The project also impressed on me the importance of standardization. As the final chapter in this book, I would like briefly to describe some basic points concerning the standardization work undertaken at these production facilities.

CLEAR PRESENTATION OF STANDARDIZATION CONCEPTS

Let us first examine some basic environmental data regarding the New England production facilities of this company.

1. *High employee turnover rate.* Employee turnover was about 2.5 to 3.0 percent per month. In addition to internal job transfers, employees changed companies to obtain higher salaries and job status.
2. *Difficulties in communication.* The production plant operated 24 hours a day, seven days a week on alternating shifts, using four work teams. The Japanese managers and locally hired supervisors worked five 8-hour shifts per week, which made it hard for them to keep in touch with all the work teams, to provide guidance, and simply to communicate.

3. *Ineffective training.* Japanese operations instructors provided training on an individual basis; however, due to the high turnover rate, as well as language and cultural differences, the training was not very effective.

In this environment, the standards themselves played the role of interpreter in communicating the operating technology. We worked out the following standardization measures for this situation:

- We established one leader for each work team, provided these leaders with a clearly defined list of duties, and planned a standardization program based on leader-centered team activities and manager-supervisor committee meetings. We also assigned the team leaders the task of training their teams in newly established standards.
- We established standards using visual management methods. To do this, we trained managers and supervisors in visual management and put them to work on standardization projects, assisted by Japanese employees who had expertise in specific standardization approaches.
- We worked to upgrade the methods, documents, and steps used for delegating work and promoted greater use of measuring instruments and other forms of mechanization to develop equipment that builds in quality at the source. To do this, we decided to increase the number of equipment maintenance employees, augment their training, and strengthen the maintenance system.

Understanding the concepts behind standardization became particularly important in this Western context. Understanding these concepts is important in Japan, too, but in my experience, American workers want to understand the reasons behind what they are asked to do.

Accordingly, we explained the concept of standardization and encouraged everyone to reflect on it whenever problems arose:

> Standards contain the information workers need to know in order to make good products reliably, easily, and quickly. These standards must be easy for workers to understand and must present the steps they should follow in doing their work. Accordingly, these standards should be easy to master, even when a worker has little experience with the task.

Some specific examples of these principles in action at the U.S. plant follow.

HOW TO MAKE STANDARDS REALLY UNDERSTOOD

1. Find familiar analogies from everyday life to explain the contents of standards. For example, we compared oil level checks in production equipment to checking the motor oil level in an automobile.
2. For response measures that operators found difficult to learn (such as setting the casting cycle or determining the shapes of jigs or tools), we included drawings in the standards that clearly show the correct measures; operators were instructed to simply follow the drawings. Although experienced Japanese managers often use a flexible approach to troubleshooting that relies on quick problem evaluation and learned responses, for this situation we felt it was much better to create simple, clearly explained standards to follow.
3. When standards failed to get the full approval and compliance of the operators, the managers' and supervisors' committee met, with an interpreter, to discuss the problem thoroughly and consider alternatives suggested by the employees. Although these proposed alternatives did not work out, we assigned committee members to assist the operators in making whatever changes and improvements were needed to enable the standards to be carried out.
4. We used photographs, figures, and charts as much as possible. For "must" items, these were posted on or near the machines and checked often. For "non-musts," we created standards manuals, kept at the group leader's desk. Shop floor training was carried out for both by types of standards.

When standards could not be built into the equipment, we tried to use the standards manual effectively by using, for example, the SN checkpoint development method described in Chapter 3.

Developing Activities to Enforce Standards

- Using the four work teams, we drew up categories for each team — including quality, case production output, and delivery factors — and devised a scoring system modeled after American football so that the teams could compete against each other. We sponsored study group and committee meetings designed to help the teams achieve higher scores. This program met with a surprisingly high degree of enthusiasm.

- We put certain people in charge of certain standards and set up an auditing system to test trainees for specific qualifications, such as those required for Ford's Q1 certification.
- We went beyond enforcing standards by setting up a suggestion system, improvement project teams, and voluntary improvement activities on holidays (with overtime pay). These efforts put people on the road toward a high level of standardization. To help expedite the implementation of suggested improvements, we also organized equipment improvement teams.

Making the Most of Employee Abilities

As the standardization activities progressed, the Japanese managers came to appreciate certain abilities of the American workers and strove to make the most of those abilities. The following are some of our observations:

- The operators generally showed strong interest in their work and were enthusiastic and diligent in their efforts to master new tasks quickly. Some people even came to work on their days off to make such efforts. They did not hesitate to ask questions, and many employees stayed overtime to discuss problems and study.
- Younger workers were not afraid to step forward and show leadership in helping the work along.
- The employees found ways to make difficult work enjoyable. They showed us many examples of American ingenuity, particularly after we established a strong organization for implementing improvements.

Revising Standards from the Operator's Perspective

The people who write standards may believe that what they write will be easily understood by operators, but they are sometimes wrong. Problems frequently arise from standards that are unclear.

Figure 10-1 shows an example from the U.S. factory; the left side shows the original standard, with the revised standard on the right. This standard concerns a mobile robot arm used in automated equipment. The factory had problems getting the robot arm to pick up and convey wheel-shaped parts correctly. Consequently, the arm sometimes pushed parts onto the floor instead of grasping them, and the equipment also suffered frequent minor stoppages due to the robot arm striking the carrier or the part; these errors showed up later as defective products and equipment breakdowns.

To study this problem, the improvement team made a videotape recording and used IE analysis to identify the key technical points in the robot arm setting procedure. Before the revision, the arm was centered as shown in the left side of the figure. But it was difficult to maintain this setting, and consequently problems occurred frequently.

Whenever a problem occurred, a supervisor went over the concepts behind the robot-setting procedure with the operator, and the operator appeared to understand them. But the setting method described in the left side of the figure was not a reliable, easy, and quick method; instead, it was ambiguous and left too much room for individual interpretation.

The revised standard drawing on the right side of the figure is easy to understand and use in checking the spaces between the robot arm's open grip and the sides of the parts carrier. It explains the setting procedure much more clearly.

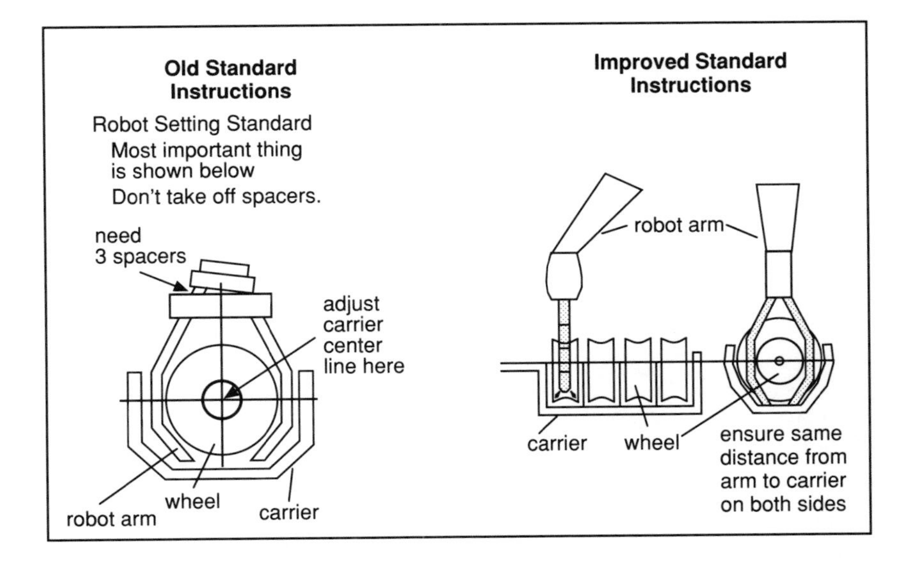

Figure 10-1. Example of Standard Revised by U.S. Affiliate Company (Robot Setting Standard)

This is an example of the good results that can be achieved by reflecting back on one of the principal concepts of standardization: revise standards so they work from the operator's standpoint. Easy, visual standardization approaches are preferable to numbers or technical instructions for accomplishing the task.

Because of the many difficulties encountered in one-on-one training, we decided to change our methods and make such training just one element of a comprehensive program of standardization activities.

Thanks to these and other standardization achievements, we raised the levels of quality, yield, and productivity, achieving our initial plant investment targets despite major changes in the product line.

The pursuit of standardization is a fundamental part of any manufacturing enterprise. No matter who your employees are or where you operate, you can make good use of the ideas presented in this book, adjusting them to suit your particular circumstances. I hope the ideas and examples presented here will be helpful as you pursue standardization at your own company.

About the Author

Shigehiro Nakamura graduated from Waseda University with a master's degree in metal engineering. He joined Hitachi Metal in 1970 and introduced and deployed companywide industrial engineering during his first eleven years with the company. Mr. Nakamura also built a pilot JIT plant as well as a pilot FA line at another plant and improved the subcontractor systems. He went on to promote VEC and CIM throughout the company.

More recently, Mr. Nakamura managed the building and startup of AAP St. Mary in the United States. He was also manager of process control/equipment control and system improvement, and in charge of the CIM project. He joined the Japan Management Association in 1991 as a Total Productivity management consultant and instructor for JMA's management school.

Index

"5M" categories, 78-79
  machines, 87
  materials, 88
  measurement, 88-89
  men and women (the Work Force), 86-87
  methods, 87
"5S" (industrial housekeeping) activities, 83, 115, 117-119
"5W1H" questions, 5-6, 109
  identifying problems via, 24, 28, 91-94

Abnormalities. *See also* Problems
  identifying, 32, 156-161
  responding to, 38, 41-44
Activities to maintain standards, 104-106
Analytical vs. practical approaches, 108-111
Automation. *See also* Equipment management; FA (factory automation)
  simplify manual methods before, 204-205
  steps toward, 146-147
Automotive assembly lines, 168-171

Bodek, Norman, ix-xii
Bottlenecks, 185, 197
Brainstorming and problem solving, 28-29
Breakdowns
  analysis using sensor devices, 156-161
  data on, 124, 131
  diagnostic instruments, 159-161
  and inspections checklists, 115-121, 130, 133, 135-138
  MTBF (mean time between failures) calculations, 125, 126
  reducing through systems and procedures, 120, 132-138
  understanding principles behind, 117-119, 122-124
  visual management and analysis of, 124-132

Case studies
  analyzing problems via, 91-96
  broken belt in NC machining center, 95-96
  cracks in cast items, 91-94
  poor die closures in casting equipment, 94-95
Categories of standards, 14-16
  by applications, 15
  by basic characteristics, 14-15
  by objects, 16

Causal-factor diagram, 80-82
Cause-and-effect diagram, 80-83
Changeover
  developing efficient single, 151-156
  quick, 60, 151-156, 217
  work-in-progress and, 232-234
Charts
  display charts, 100-101, 126
  evaluation chart for improvement levels, 99
  idea evaluation chart, 101, 103
  operator-machine charts, 179-181
  PERT, 190-197
  problem-analysis table, 99, 100
  process charts, 181-183
  as QC tools, 112-113
  skills achievement chart, 70-71
Checklists, inspection, 115-121, 130, 133, 135-138
Chip management, 146-151
Cleaning, 121, 122
Color coding and checking procedures, 33-34, 126-128, 132
Competition and distinctive quality, 86-89
Complaints, responding to, 242-245
Computer integrated manufacturing (CIM), 9-12, 63, 66, 171-172, 208
Computerization
  benefits of, 66
  caution points regarding, 215-217, 218-223
  lead-time shortening for information-based processes, 215-217
  of machines, 87
Continuous improvement, ix, 8-9. *See also* QC (quality control); Quality
  adopting in the United States, xiii-xiv
  improvement plans, choosing best, 101-103, 106
  and Plan-Do-Check cycles. *See* Plan-Do-Check cycles
Control points in standards manuals, 29-32
Conventional approach, errors in, 1-5
Countermeasures
  prioritizing, 129-130
  for problems, 99, 110, 125, 145
  for rush orders, 226-228
Customer needs, 83-86
Cutting oil, 150-151
Cyclegraph and chronocyclegraph, 168

da Vinci, Leonardo, 165
Data
  extracting needed, 63-66
  on breakdowns, 124, 131
Decision making overview, 239
Defective products, 8. *See also* Zero defects
Design stage of standardization, 58-59
Diagnostic instruments, 159-161
Display charts, 100-101, 126
Double-checking standards, 39-40

Economic lot sizes, 232-233
ECRS (eliminate, combine, rearrange, simplify) approach, 47
Employees
  employee-to-employee training, 67-70
  making most of abilities, 256
  turnover, 253
Errors in approach to standardization, 1-5
Equipment management breakdowns. *See* Breakdowns
  capacity utilizations, 124, 125
  daily equipment checks, 106, 117, 118, 133
  deterioration, 119, 123
  inspection checklists, 115-121, 130, 133, 135-138

Equipment management breakdowns *(cont.)*
maintenance of tools and equipment, 116, 120-122
new-equipment installation, 139-140
priorities and plans, establishing, 129-130
and project teams, 129-130
shutdown maintenance, 140-143
unassisted equipment operation, 144-151
Evaluation chart for improvement levels, 99

FA (factory automation), 151-161, 171-172, 235-237
breakdown analysis using sensor devices, 156-161
developing efficient single changeover, 151-156
standards and, 12, 107
Factory management, 19-23
Factory standard, 104
Feigenbaum, Armand, *Total Quality Control*, 77
Fishbone diagram. *See* Cause-and-effect diagram
Flowcharts, 57, 189
Flow of work. *See also* Production; Production management
improvement in, 189, 234-237
standards related to, 6-7
FMS (flexible manufacturing systems), 151-161, 171-172, 235-237
breakdown analysis using sensor devices, 156-161
developing efficient single changeover, 151-156
kaizen for smoother production flow, 228-237
Ford, Henry, 9, 11
Functional quality, 78

GCM (greatest common measure) part, 58
Gilbreth, Frank B. and Lillian M., 166-168
GT (group technology), 12, 57, 58-61

IE (industrial engineering) methods, 45-50
Improvement plans, choosing best, 101-103, 106
Improving standards, 3, 13-17, 37-41, 45-52
Inspections
daily equipment checks, 106, 117, 118, 133
in-process, 40-41
checklists, 115-121, 130, 133, 135-138
International Standards Organization (ISO), 9
Investment plans, 246-252
Ishikawa diagram. *See* Cause-and-effect diagram
Item sheet, 32, 33

Japan Management Association, ix, xi
Japanese continuous improvement. *See* Continuous improvement
Japanese Standards Association, 13
Job-shop system, 212
Just-in-time production (JIT), xvii, 61, 208, 217

Kaizen, 138, 228-237
Kanban system, 61, 63, 64, 127-128, 171
Know-how, 96-107
good and bad, 96-98
making standards that emphasize, 98-107

Lead time, shortening production, 208-211, 215-218
Load leveling. *See* Production, leveled

Machines, 87
Maintaining standards, 89-90
Maintenance
  cards, color-coded, 126-128, 132
  and equipment inspection checklists, 120-122, 130, 133, 135-143
  logs, 122
  preventative/productive (PM), 50
  schedule, using PERT, 141, 143
Managers. *See also* Operation management; Production management
  role in work site, 146
  and standards, 72-75
  training by, 70-71
Manuals, 56-57. *See also* Standards manuals
Manufacturing
  history of, 9, 11
  standards, 19-23. *See also* Technical standards
  technologies, clarifying, 99-101
Materials, 88
Matsushita Washing Machine plant, xi
Measurement, 88-89
Mechanization, categories of, 204
Meeting preparation, 239-242
Men and women (the Work Force), 86-87
Methods. *See* work methods
Micromotion study, 168
Mistake proofing. *See* Poka-yoke (mistake-proofing)
Monitoring for minor stoppages, 144-146
Motion study, 167-169
MRP (materials requirement planning), 220, 223
MTBF (mean time between failures) calculations, 125, 126
Musts and preferences, 29-32, 101, 117, 255

Nakaigawa, Masakazu, 123-124
Nakamura, Shigehiro, xv-xvii
New-employee training, 25-26, 66-72
New-equipment installation, 139-140
Nippon Steel's Kimitsu Works, 157

O'Brien, Mark, Forward, xiii-xiv
Ono, Shigeru, 9
Operation instructions, 57-63
Operation management
  from Ford system to Toyota production system, 168-171
  Gilbreth's work, 166-168
  increasing effective improvement methods, 173-174
  overview, 163
  problem-consciousness, 3-5, 183-187
  Taylor's work, 9, 163-166
Operation standards. *See also* Standards
  daily equipment checks, 106, 117, 118, 133
  determining methods and levels of, 172-175
  developing and displaying, 171-183
  nonexistent or too lenient, 6-9
  operator-machine charts, 179-181
  process charts, 181-183
  standard time system, 187-197
  steps in making, 175-177
  stopwatch method, 175-179
  time-based analysis, 172-173
Operator-machine charts, 179-181
Operators
  as researchers, 107, 121, 145
  and their equipment, 117

P-D-C cycle. *See* Plan-Do-Check cycle
PERT (program evaluation and review technique), 190-197, 141, 143, 176
  applying in furniture-moving operations, 190-197
Plan-Do-Check cycles
  improving standards via, 19-20
  standardization of know-how, 96, 98
  standardization of problem analysis, 89-91
  and production management, 208-209, 220
Poka-yoke (mistake-proofing)
  devices, xi, 4, 17
  methods, 51-52, 88-89, 111-114
Preferences and musts, 29-32, 101, 117, 255
Problem finding method. *See also* SN method
  checklist, 92-93, 110
  idea evaluation chart, 101, 103
  listing problem points, 98
  problem-analysis table, 99, 100
  problem points, listing, 98
  studying countermeasures for problems, 99, 110, 125, 129-130, 145
Problems. *See also* Abnormalities; Problem finding method; Work-site problems
  analysis of, 89-96
  and causal-factor diagrams, 82
  and cause-and-effect diagrams, 82-83
  communicating via visual management, 71-72
  consciousness in standardization, 3-5, 183-187
  decision making and, 239
  problem-solving sequence, 24-29, 223, 226
  studying countermeasures for, 99, 110, 125, 129-130, 145
Process
  charts, 181-183
  operations and inspections. *See* inspections
  progress report, 32, 33
  standards, 14-15, 54-56
  time (PT), 50
Product line standardization, 59-61
Production
  instructions, 63
  leveled, 60, 61-63, 171, 212-214
  just-in-time production (JIT), xvii, 61, 208, 217
Production management
  daily, 218-223
  economic lot sizes, 232-233
  improving links between processes, 229-232
  and kanban system, 61, 63, 64, 127-128, 171
  leveled production, 60, 61-63, 171, 212-214
  objectives, 207-208
  overall flow improvement, 234-237
  rush orders, 223-228
  sales approach, 207-209
  shortening lead time, 208-211, 215-218
  small-lot production, 212
  and standard time systems 187-197
  technology and standards, 5-9
  turnover approach, 207-211
  types of systems, 229
  using visual management, 71-74
  work-in-progress, 233-234
Project teams, 129-130, 239-242

QC (quality control), x, xvii
  current standards level, 50

QC (quality control) *(cont.)*
problem solving via, 45, 47
tools and their uses, 111-114
QCD (quality, cost and delivery), 2-3
and new standards sheets, 104
and standardization, 11-12, 91
Quality
appraisal factors, 86, 87
and causal-factor diagram, 80-82
and cause-and-effect diagram, 80-83
and customer needs, 83-86
creating distinctive, 86-89
definition of, 77-78
factors within "5M" categories, 78-79, 86-89
functional, 78
improvement, 78-83
true and substitute characteristics of, 83-84
and zero defects, 108-111
Quality assurance checkpoints, 32, 133
Quality control. *See* QC (quality control)
Quality standards. *See* Standards

Results/target ratio, 6, 8
Result time (RT), 220
Rush orders, 223-228

Safety awareness training (SAT), 50, 133, 136, 137
Safety calendar, 140
Safety, health and environmental protection, 29
Sales approach to production management, 207-209
Shimizu, Hiroshi, xi
Shingo, Shigeo, 97-98, 111, 146, 152, 233
Shutdown maintenance, 140-143
improving links between processes, 229-232
Single changeover, 151-156
Small-lot production, 212
SN method. *See also* Problem finding method
approach to flow improvement, 234-237
case study using, 91-94, 110, 114
and customer complaints, 245
definition of, xvii
and equipment capacity utilizations, 124, 125
evaluation of current standards, 49-52
operators as researchers, 107, 121, 145
and overall flow improvement, 234-237
and plant investment plans, 246-252
problem finding checklist, 92-93, 110
working toward factory automation, 107
Specialists, 70-71
Standard time (ST) system, 164, 187-197, 220
Standardization
automating and upgrading, 197-205
basic steps in, 90-91
for breakdown analysis, 122-132
and continuous improvement, ix
conventional approach to, errors in, 1-5
design stage, 58-59
for the FA/CIM era, 9-12
and IE methods, 47-49
key concepts of, 2-3
of know-how, 96-107
making standards really understood, 255-258
of meeting preparation, 239-242
for new-employee training, 66-72
of operation instructions, 57-63

Standardization *(cont.)*
overview, ix
of problem analysis, 89-96
presenting concepts clearly, 253-254
problem-consciousness before, 3-5, 183-187
product line, 59-61
and QCD (quality, cost and delivery), 11-12, 91
responding to complaints, 242-245
response to abnormalities, 32, 38, 41-44
sequence for problem solving, 26-29
unassisted operation, 144-151
from user's perspective, 32-36
zero-defect improvements, 46-47
Standards
activities to maintain, 104-106
advantages of user-friendly, 36
basic characteristics of, 14
corporate chain of responsibility and, 74-75
and corporate health, 9, 10
and current level of production technology, 5-9
categories of, 14-16
creating explicit, 103-104
definition of, xvi, 13-14
determining current and higher levels, 49-52
developing activities to enforce, 255-256
double-checking, 39-40
formula for establishing, 15
importance of maintaining, 89-90
improving, 3, 13-17, 37-41, 45-52
making and checking, 37-41
and managers, 72-75
from manufacturing perspective, 19-23
object-based categorization of, 16
operation. *See* Operation standard
from operator's perspective, 256-258
raising level of, 45-52
and results/target ratio, 6, 8
technical and process, 14-15, 54-56
and technological advances, 3
troubleshooting, 43-44
and work site problems, 23-29
workplace applications and, 15
written, 53-57
Standards sheets
creating new, 37-39, 104
dealing with poorly prepared, 37-38
evaluating, 38
making easy-to-use, 38-39, 54, 104
posting, 106
Standards manuals, 3-5, 29-32
must and preference items, 30, 101, 117, 255
Stopwatch method, 175-179
Storage points, 218, 231

Talbot, Bruce, xi
Taylor, Fredrick W., 9, 163-166
The Principles of Scientific Management, 164
Technical principles and quality, 78-81
Technical standards, 14-15, 54-56
clarifying manufacturing technologies, 99-101
as must items, 32, 101, 117, 255
Technological advances and standards, 3
Therbligs, 167-168
Time study, 152, 154-155, 163-166
Tools, maintenance of, 116
Total productive maintenance, x
TPM (total productive maintenance), 115, 119, 121
Training
employee-to-employee, 66, 67-70
new-employee, 25-26, 66-72

Training *(cont.)*
  skills achievement chart, 70-71
  by specialists or managers, 70-71
  by visual management, 71-72
Troubleshooting, 43-44, 57
Turnover approach to production
    management, 207-211

Uchiyama, Kazuya, xi
Unassisted equipment operation,
    144-151
User-friendly standards, 2, 36

VE (value engineering), 45, 47, 153, 199
  and functional family tree
    development, 199, 203
Visual aids, xi
Visual management, 124-132
  color-coded maintenance cards,
    126-128, 132
  and daily production management,
    220, 222
  display charts, 100-101, 126
  establishing priorities and plans,
    129-130
  following up on effects, 130-132
  training by, 71-72

Waste, 214
Work-in-progress, 232-234
Work instructions, 32, 33, 63, 65. *See also*
    Standards
Work methods, 87, 163
Work-site problems, 23-29
  defining, 24
  problem-solving sequence, 24-26
Written standards, 31, 53-57

Yanagida, Norio, Fear at Mach 1, 28

Zero defects, 108-111
  analytical vs. practical approaches,
    108-111
  improving standards to achieve, 46-47
  QC tools and, 111
  using mistake-proofing methods,
    51-52, 88-89, 111-114